Electrocatalysts and Advanced Materials for Sustainable Energy Storage

Edited by

**Kalathiparambil Rajendra Pai Sunajadevi[1],
Dephan Pinheiro[1] and Mothi Krishna Mohan[2]**

[1]Department of Chemistry, Christ University, Bangalore, India

[2]Department of Sciences and Humanities, Christ University, Bangalore, India

Copyright © 2025 by the authors

Published by **Materials Research Forum LLC**
Millersville, PA 17551, USA

Published as part of the book series
Materials Research Foundations
Volume 182 (2025)
ISSN 2471-8890 (Print)
ISSN 2471-8904 (Online)

Print ISBN 978-1-64490-378-0
eBook ISBN 978-1-64490-379-7

Distributed worldwide by

Materials Research Forum LLC
105 Springdale Lane
Millersville, PA 17551
USA
https://mrforum.com

Manufactured in the United States of America
10 9 8 7 6 5 4 3 2 1

Table of Contents

Preface

Preface

As the world moves steadily toward renewable energy, the need for efficient, reliable, and environmentally friendly energy storage has never been greater. Solar panels and wind turbines are transforming how we generate power, but we are aware of their intermittent nature, and without advanced methods to store this energy, their full potential cannot be realised. This is where electrocatalytic materials come into play—helping us store energy more efficiently, sustainably, and on a scale that meets global demand.

This book, *Electrocatalysts and Advanced Materials for Sustainable Energy Storage*, brings together some of the most exciting developments in this fast-evolving field. It offers readers a broad yet detailed exploration of how different materials—ranging from metal-organic frameworks (MOFs) and MXenes to carbon nanotubes, graphene, biomass-derived catalysts, and transition metal oxides—are being designed, synthesised, and optimised to power the energy systems of the future.

We have structured this book to serve both as an introduction for newcomers and a valuable reference for seasoned researchers and professionals. Each chapter focuses on a specific class of material or a key challenge in the field, from enhancing the performance of supercapacitors to improving the stability of energy storage devices. Along the way, we also highlight environmentally conscious approaches—such as the use of renewable biomass-based catalysts—that are helping to make the entire process more sustainable. Our goal is to provide not just technical insights but also to spark new ideas, collaborations, and innovations. Energy storage is a challenge that touches every aspect of modern life, from powering electric vehicles to stabilising national power grids. The materials and technologies discussed in this book have the potential to make a meaningful difference in tackling some of the biggest energy challenges we face today.

We hope this book will be a useful and an inspiring resource for scientists, engineers, students, and anyone passionate about building a cleaner, more sustainable energy future. The journey toward that future is still unfolding, and the materials we choose to work with will shape the path ahead.

Electrocatalysts and Advanced Materials for Sustainable Energy Storage   Materials Research Forum LLC
Materials Research Foundations 182 (2025) 1-11   https://doi.org/10.21741/9781644903797-1

Chapter 1

Electrocatalysts towards Sustainable Energy Storage Applications

ARUN Varghese[1,a*], Kalathiparambil Rajendra Pai SUNAJADEVI[2,b*], SONY J. Chundattu[3,c]

[1]Department of Sciences, Alliance School of Sciences, Alliance University, Central Campus, Anekel, Bangalore -562106, India

[2]Department of Chemistry, Christ University, Hosur Road, Bangalore-560029, India

[3]Department of Sciences and Humanities, School of Engineering and Technology, Christ University, Bangalore, 560074, India

[a]arunvarghese94@gmail.com, [b]sunajadevi.kr@christuniversity.in, [c]frsony@christuniversity.in

Abstract

Effective storage solutions are required as the world moves toward sustainable energy to handle the sporadic nature of renewable energy sources like solar and wind. An essential component of this effort is electrocatalysts, which can speed up important electrochemical reactions to improve the efficiency of energy storage devices like fuel cells, batteries, and supercapacitors. By exploring the basic ideas of electrocatalysis, this introductory chapter clarifies how these materials support energy conversion processes. It offers a thorough analysis of different electrocatalytic materials, including metal-organic frameworks, carbon-based compounds, and catalysts based on transition metals, going into their special qualities and uses in energy storage systems. Along with discussing current issues, including cost-effectiveness, scalability, and stability, the chapter also identifies new developments and potential avenues for electrocatalyst research. This introduction seeks to provide readers with a thorough grasp of the vital role electrocatalysts play in developing sustainable energy storage systems, laying the groundwork for the next chapters.

Keywords

Sustainability, Energy Storage, Electrocatalysts, Supercapacitors

Contents

Introduction to Electrocatalyst for Sustainability

At a time of growing environmental concerns, fast industrialization, and an increasing need for clean and efficient energy, sustainable electrocatalysis has become more of a mandate than a choice. The Sustainable Development Goals (SDGs) of the United Nations, especially Clean Water and Sanitation (SDG 6), Affordable and Clean Energy (SDG 7), and Industry, Innovation, and Infrastructure (SDG 9), are greatly advanced by it, and it is at the center of a worldwide movement for green technologies [1–4]. When used responsibly, electrocatalysis, the term for the process by which catalysts speed up electrochemical processes at electrode surfaces, allows renewable electricity to be converted into chemical energy that can be stored with little harm to the environment. From producing clean water via electrochemical purification to supplying energy for future cities via hydrogen fuel cells and supercapacitors, sustainable electrocatalysis provides a way to balance environmental responsibility with technological progress [5,6]. Historically, the development of electrocatalysis dates back to the early studies on electrolysis and fuel cells in the 19th and 20th centuries. Initially focused on precious metal catalysts like platinum, the field has evolved to embrace more abundant and eco-friendly materials such as transition metal oxides, conductive polymers, carbon-based nanostructures, and metal-organic frameworks (MOFs). The evolution of sustainable electrocatalysts has been driven by the dual needs of performance and sustainability, delivering high catalytic activity while reducing reliance on scarce, toxic, or expensive elements [7,8].

In the context of energy storage, sustainable electrocatalysis is essential for improving the efficiency, durability, and scalability of advanced systems like lithium-ion batteries, metal-air batteries, and electrochemical capacitors [9]. These systems rely on fast and reversible redox reactions at the electrode-electrolyte interface, processes which are heavily influenced by the nature of the electrocatalyst. For instance, in water splitting, a crucial strategy for green hydrogen production, electrocatalysts govern the sluggish oxygen evolution and hydrogen evolution reactions, determining the energy input and overall efficiency [10,11]. Similarly, in supercapacitors, pseudocapacitive materials enhanced with electrocatalysts can significantly improve energy density while maintaining high power performance [12]. Beyond energy, electrocatalysis also plays a major role in environmental remediation, such as the degradation of pollutants through advanced oxidation processes and adsorption-enhanced electrochemical techniques [13]. Researchers are now exploring multifunctional catalysts capable of addressing both energy and environmental challenges in one platform. As we continue to seek solutions that are not only effective but also environmentally and economically sustainable, the development of novel electrocatalytic materials is poised to define the future of clean technology.

To truly align with sustainable development, future electrocatalysts must fulfill several criteria, including being derived from earth-abundant materials, exhibiting long-term operational stability, possessing high efficiency at low overpotentials, and being synthesised through eco-friendly and scalable methods. Thus, sustainable electrocatalysis does not simply refer to the application of catalysts in green technologies; it encapsulates a broader vision of science that contributes to a circular, low-carbon, and equitable future. As we go along this chapter, we will explore how these catalysts function, their classification, and their diverse applications in energy storage, paving the way for a more resilient and energy-secure world.

Fundamental Principles of Electrocatalysis

Electrocatalysis forms the backbone of many modern energy storage technologies, including water electrolyzers, metal-air batteries, supercapacitors, and various hybrid systems. In such devices, the efficiency, durability, and power output are largely determined by the ability of electrocatalysts to facilitate electrochemical reactions at the electrode-electrolyte interface. A comprehensive understanding of the fundamental principles of electrocatalysis, encompassing electrode processes, reaction mechanisms, and activity-influencing factors, is critical for the rational design of next-generation energy storage systems [14]. In electrochemical energy storage devices, the key processes often involve the interconversion of electrical and chemical energy during charging and discharging cycles. These processes are governed by redox reactions occurring at the electrodes, with ion migration through the electrolyte and electron flow through an external circuit. For example, in water-splitting cells, the hydrogen evolution reaction (HER) and oxygen evolution reaction (OER) are central to energy conversion, while in metal-air batteries, the oxygen reduction reaction (ORR) plays a dominant role during discharge. The kinetics of these reactions, essentially how quickly they proceed, have a direct impact on how efficiently energy can be stored or extracted [15]. Real-world electrochemical reactions often require an extra driving force beyond the thermodynamic potential, known as overpotential, to overcome activation energy barriers. Reducing this overpotential is a critical challenge in the design of a catalyst, as it enhances overall energy efficiency. The Butler–Volmer equation provides a mathematical description of the relationship between current density and overpotential, offering valuable insight into the kinetic behavior of catalysts. A catalyst with a high exchange current density and a low Tafel slope is especially desirable for overall water splitting studies [16]. Faster charge transfer and superior electrocatalytic performance, an essential feature for high-power applications, are essential for supercapacitors and rapidly charging energy storage devices [17].

Equally important is the mechanistic understanding of how electrocatalytic reactions unfold. These reactions typically involve multiple steps, such as adsorption of reactants, electron or proton transfer, intermediate formation, bond rearrangement, and product desorption [18,19]. For instance, HER usually follows either the Volmer–Heyrovsky or Volmer–Tafel pathway, with each involving adsorbed hydrogen intermediates. On the other hand, OER, a more complex four-electron process, often proceeds through energetically demanding intermediates like OOH* [20]. Such mechanistic knowledge is critical for addressing performance bottlenecks in energy storage applications, particularly where slow kinetics limit device efficiency. Identifying the rate-limiting steps and tailoring the catalyst's surface to optimize the binding energy of intermediates can significantly enhance overall performance. Designing catalysts that strike a balance between stabilizing critical intermediates and ensuring their timely desorption is key to achieving both high reaction rates and operational longevity [21].

The electrocatalytic activity and stability of materials are governed by a blend of intrinsic and extrinsic factors. Intrinsic factors include the electronic structure of the catalyst, which dictates the density of states near the Fermi level and, consequently, the ease of electron transfer. Catalysts with optimized electronic configurations can facilitate lower activation energies, thereby improving reaction kinetics. Additionally, surface area and morphology play a crucial role; nanostructured catalysts with high surface-to-volume ratios offer more accessible active sites and improved mass and electron transport, advantages that are particularly beneficial in supercapacitors and hybrid energy devices [22]. Crystallographic orientation and structural defects can also enhance catalytic activity by exposing high-energy facets or creating active defect sites. Among the extrinsic factors, the electrolyte composition and pH can drastically affect ion mobility and catalyst stability. Alkaline electrolytes are often preferred in water-splitting and metal-air systems due to their favorable ionic conductivity and reduced corrosion of non-noble catalysts. Operational parameters such as temperature, applied voltage, and pressure can further influence the kinetics and long-term performance of catalysts. For instance, elevated temperatures may enhance ion mobility but could accelerate catalyst degradation if thermal stability is not ensured. Furthermore, catalyst-support interactions, especially in composite materials, play a crucial role in performance optimization. The synergy between metal nanoparticles and conductive supports like doped carbon or metal oxides often results in improved electron transport, mechanical stability, and resistance to degradation [16,23].

For energy storage applications to be practically viable, electrocatalysts must not only exhibit high activity but also maintain their structural integrity and functional efficiency over extended operational cycles. Strategies such as encapsulating active materials, introducing dopants or forming stable alloys, and employing robust support matrices are being actively pursued to enhance durability and prevent performance degradation. Ultimately, mastering the fundamental principles of electrocatalysis provides the scientific foundation for designing advanced materials that can address the core challenges of cost, efficiency, and longevity in sustainable energy storage technologies. Whether it's accelerating oxygen evolution in an electrolyzer, boosting redox kinetics in a metal-air battery, or enhancing charge-discharge cycling in a supercapacitor, each electrocatalytic parameter plays a vital role in shaping the future of clean energy systems.

Types of Electrocatalytic Materials

In the transition to a sustainable energy future, the performance of energy storage systems such as supercapacitors, metal-air batteries, and water electrolyzers hinges heavily on the nature of the electrocatalysts involved. Electrocatalytic materials play a vital role in reducing energy losses, accelerating reaction kinetics, and improving overall efficiency and durability. The following discussion outlines four major categories of electrocatalytic materials: metal-based, carbon-based, metal-organic frameworks (MOFs), and polymer-based systems, with a special emphasis on their applications in energy storage technologies.

Metal-based catalysts are among the most widely used electrocatalytic materials in energy storage systems due to their robust activity and well-defined reaction mechanisms. In devices such as water electrolyzers and rechargeable metal-air batteries, these catalysts are often employed for the HER, OER, or ORR, all of which are central to efficient energy conversion and storage. Platinum-group metals like Pt, Ir, and Ru are benchmark catalysts due to their superior performance [24]. However, their scarcity and high cost pose significant barriers to scalability. As a result, extensive research has turned toward earth-abundant transition metals such as Ni, Co, Fe, and Mn [25,26]. These metals, particularly in the form of oxides, phosphides, or sulfides, have demonstrated significant potential in alkaline media [27].

Carbonaceous materials are indispensable in the design of energy storage devices due to their excellent electrical conductivity, lightweight nature, and mechanical stability [22]. While pure carbon materials like activated carbon or graphite offer a high surface area, their intrinsic catalytic activity is limited [28]. However, chemical modifications such as heteroatom doping, like nitrogen and sulfur, or hybridization with metal nanoparticles have transformed these materials into potent electrocatalysts for various energy-related reactions [29]. In supercapacitors, doped graphene and carbon nanotubes (CNTs) serve dual roles as charge storage materials and as electrocatalytic supports, enhancing both capacitance and reaction kinetics [30]. Furthermore, the structural tunability of carbon materials enables their integration into flexible and wearable energy storage devices [31].

MOFs are emerging as next-generation materials in the field of electrocatalysis, particularly for energy storage applications. Their high porosity, customizable metal centers, and functional organic linkers make them excellent candidates for controlled catalyst design [32]. Although pristine MOFs often suffer from poor conductivity, their derived forms, typically obtained via pyrolysis, have shown excellent performance. Pyrolyzed MOFs can yield metal-doped porous carbon composites with embedded nanoparticles, which are ideal for electrochemical energy storage. In supercapacitors, these materials offer a high surface area for double-layer formation and pseudo-capacitive behavior. The modularity of MOFs also allows the development of bifunctional catalysts, serving multiple roles in integrated energy storage systems [33,34].

Conducting polymers such as polyaniline (PANI), polypyrrole, PEDOT, and polythiophene are gaining increasing attention for their unique combination of conductivity, flexibility, and redox activity. These materials are especially useful in applications where lightweight, mechanical flexibility, and low-cost processing are important, such as in flexible supercapacitors and hybrid batteries [22,35]. In energy storage, polymer-based electrocatalysts serve both as active charge storage materials and as supportive frameworks for hosting metal or metal oxide nanoparticles. Their inherent redox properties enable them to store charge efficiently, while their compatibility with various nanomaterials allows them to function in composite electrodes with enhanced electrocatalytic activity. For instance, PANI-metal oxide composites have demonstrated excellent pseudocapacitance and long-term cycling stability in hybrid supercapacitors. Moreover, the functional groups in polymers can interact with the electrolyte species or facilitate electron/proton transport, further improving device efficiency [36].

The choice and design of electrocatalytic materials are key determinants in the performance, cost, and scalability of sustainable energy storage systems. Whether through enhancing reaction kinetics, increasing energy density, or improving cycling stability, these materials form the foundation upon which next-generation energy solutions are being built.

Electrocatalysts in Energy Storage Applications

Batteries, particularly next-generation variants such as metal-air, lithium-sulfur, and sodium-ion batteries, rely heavily on electrocatalysts to overcome kinetic limitations associated with charge/discharge processes. In metal-air batteries, such as Zn–air or Li-air systems, the oxygen reduction reaction (ORR) during discharge and oxygen evolution reaction (OER) during charging are critical to overall performance. Both reactions suffer from sluggish kinetics and require highly active, stable electrocatalysts to ensure energy efficiency and longevity. Transition metal oxides, perovskites, and carbon-supported metal nanoparticles have shown promise as bifunctional catalysts, capable of facilitating both ORR and OER in these systems [37,38]. In lithium-sulfur batteries, polysulfide shuttle effects and sluggish redox kinetics can be mitigated by electrocatalysts that promote the reversible conversion of sulfur species and suppress intermediate dissolution [39]. Metal sulfides, metal-organic frameworks (MOFs), and doped carbon materials have demonstrated excellent catalytic activity toward sulfur redox reactions, enhancing capacity retention and cycling life. Sodium-ion batteries, which are gaining attention due to their low cost and resource abundance, also benefit from electrocatalyst integration to improve sodium-ion diffusion and electrode reaction kinetics [40]. The incorporation of catalytic components into the electrode design, such as layered metal oxides or heteroatom-doped carbons, has resulted in enhanced reaction reversibility and lower internal resistance, thereby improving rate capability and energy density. The electrocatalysts in battery systems are crucial not only for boosting the reaction rate and energy output but also for enabling the stable operation of electrodes under repeated cycling and high current loads.

Supercapacitors, known for their exceptional power density and rapid charge-discharge capability, are another domain where electrocatalysts significantly enhance performance. Unlike batteries, which rely on faradaic redox reactions for energy storage, supercapacitors utilize either electrochemical double-layer capacitance (EDLC) or pseudocapacitance mechanisms [35,41]. Electrocatalysts contribute mainly in the latter, where surface or near-surface redox reactions dominate charge storage. Pseudocapacitive materials, such as transition metal oxides like RuO_2, MnO_2, and $NiCo_2O_4$, conducting polymers like polyaniline and polypyrrole, and MXenes, benefit from catalytic enhancement to increase charge storage efficiency and redox kinetics [42–44]. The integration of electrocatalytically active sites into these materials improves electron and ion transport, reduces charge-transfer resistance, and enables higher capacitance with stable cycling behavior. Additionally, hybrid supercapacitors, which combine battery-type faradaic electrodes with capacitive ones, rely on electrocatalysts to bridge the gap between high energy density and high power density. In such systems, the positive electrode often contains transition metal-based electrocatalysts to drive reversible redox reactions, while the negative electrode stores charge via double-layer mechanisms. Here, the performance of the device hinges on the balance of kinetics between both electrodes, a role where electrocatalyst design becomes critical. Furthermore, the scalability and economic viability of supercapacitors for commercial applications necessitate the use of cost-effective, earth-abundant electrocatalysts that can deliver high performance with minimal degradation over extended cycles. Ongoing research into nano structuring, heteroatom doping, and synergistic composite architectures is paving the way for next-generation supercapacitor technologies that leverage the full potential of electrocatalysis [45,46].

The electrocatalysts serve as the performance-driving engine in both batteries and supercapacitors, determining how efficiently and reliably these devices can store and deliver energy. By accelerating charge-transfer processes, stabilizing intermediates, and minimizing energy losses, electrocatalysts not only enhance the electrochemical performance but also expand the practical applicability of energy storage systems. Continued research in electrocatalyst development, through material innovation, surface engineering, and mechanistic understanding, will remain at the forefront of advancing sustainable and scalable energy storage solutions.

Conclusions and Future Perspectives

The pursuit of high-efficiency, durable, and cost-effective energy storage technologies has brought electrocatalysis to the forefront of modern electrochemical research. This chapter has comprehensively explored the fundamental principles of electrocatalysis, highlighting its pivotal role in enhancing the performance of batteries and supercapacitors. By examining key electrode processes, reaction mechanisms, and factors influencing catalytic activity, we have emphasized how electrocatalysts govern crucial aspects such as energy conversion efficiency, charge-discharge kinetics, and long-term stability. The discussion of diverse electrocatalytic materials, including metal-based catalysts, carbon-based structures, MOFs, and polymer-based systems, further illustrates the versatility and potential of these materials in addressing the demands of advanced energy storage platforms. Looking ahead, several challenges and opportunities define the roadmap for future research in this field. Innovations in material design, especially the development of low-cost, earth-abundant, and environmentally benign catalysts, are essential for large-scale adoption. A deeper mechanistic understanding through advanced characterization techniques will enhance our ability to tailor catalysts for specific reactions. Moreover, ensuring scalability, mechanical robustness, and compatibility with industrial fabrication processes remains a critical challenge to overcome. The emergence of multifunctional and hybrid systems offers exciting prospects for integrating electrocatalysts into next-generation storage devices. Finally, environmental sustainability and lifecycle assessments must guide material selection and system design to ensure long-term viability. As energy systems continue shifting towards clean and renewable sources, electrocatalysis will remain a foundational pillar in the development of efficient, resilient, and sustainable energy storage technologies.

Acknowledgments

The authors are grateful to Alliance University, Bangalore, and Christ University, Bangalore, for the facilities and encouragement.

References

[1] P. Katila, C.J. Pierce Colfer, W. de Jong, G. Galloway, P. Pacheco, G. Winkel, eds., Sustainable Development Goals: Their Impacts on Forests and People, Cambridge University Press, 2019. https://doi.org/10.1017/9781108765015

[2] L. Wang, Y. Zhang, L. Chen, H. Xu, Y. Xiong, 2D Polymers as Emerging Materials for Photocatalytic Overall Water Splitting, Adv. Mater. 30 (2018) 1–12. https://doi.org/10.1002/adma.201801955

[3] A. Varghese, S. Devi K R, Tailoring a Multifunctional PEDOT/Co 3 O 4 -CeO 2 Composite for Sustainable Energy Applications, Adv. Sustain. Syst. 2300575 (2024) 1–11. https://doi.org/10.1002/adsu.202300575

[4] R.B. Swain, A Critical Analysis of the Sustainable Development Goals, in: 2018: pp. 341–355. https://doi.org/10.1007/978-3-319-63007-6_20

[5] T. Hák, S. Janoušková, B. Moldan, Sustainable Development Goals: A need for relevant indicators, Ecol. Indic. 60 (2016) 565–573. https://doi.org/10.1016/j.ecolind.2015.08.003

[6] W. Schramade, Investing in the UN Sustainable Development Goals: Opportunities for Companies and Investors, J. Appl. Corp. Financ. 29 (2017) 87–99. https://doi.org/10.1111/jacf.12236

[7] F.-Y. Chen, Z.-Y. Wu, Z. Adler, H. Wang, Stability challenges of electrocatalytic oxygen evolution reaction: From mechanistic understanding to reactor design, Joule. 5 (2021) 1704–1731. https://doi.org/10.1016/j.joule.2021.05.005

[8] F.R. Rangel-Olivares, E.M. Arce-Estrada, R. Cabrera-Sierra, Synthesis and Characterization of Polyaniline-Based Polymer Nanocomposites as Anti-Corrosion Coatings, Coatings. 11 (2021) 653. https://doi.org/10.3390/coatings11060653

[9] J. Xu, K. Wang, S. Zu, B. Han, Z. Wei, Hierarchical Nanocomposites of Polyaniline Nanowire Arrays on Graphene Oxide Sheets with Synergistic Effect for Energy Storage, ACS Nano. 4 (2010) 5019–5026. https://doi.org/10.1021/nn1006539

[10]   R. Anand, A.S. Nissimagoudar, M. Umer, M. Ha, M. Zafari, S. Umer, G. Lee, K.S. Kim, Late Transition Metal Doped MXenes Showing Superb Bifunctional Electrocatalytic Activities for Water Splitting via Distinctive Mechanistic Pathways, Adv. Energy Mater. 11 (2021). https://doi.org/10.1002/aenm.202102388

[11]   R. Djara, M.-A. Lacour, A. Merzouki, J. Cambedouzou, D. Cornu, S. Tingry, Y. Holade, Iridium and Ruthenium Modified Polyaniline Polymer Leads to Nanostructured Electrocatalysts with High Performance Regarding Water Splitting, Polymers (Basel). 13 (2021) 190. https://doi.org/10.3390/polym13020190

[12]   J. Zhang, R. Huang, Z. Dong, H. Lin, S. Han, An illumination-assisted supercapacitor of rice-like CuO nanosheet coated flexible carbon fiber, Electrochim. Acta. (2022) 140789. https://doi.org/10.1016/j.electacta.2022.140789

[13]   T. Jayaraman, A.P. Murthy, V. Elakkiya, S. Chandrasekaran, P. Nithyadharseni, Z. Khan, R.A. Senthil, R. Shanker, M. Raghavender, P. Kuppusami, M. Jagannathan, M. Ashokkumar, Recent development on carbon based heterostructures for their applications in energy and environment: A review, J. Ind. Eng. Chem. 64 (2018) 16–59. https://doi.org/10.1016/j.jiec.2018.02.029

[14]   A.G. Olabi, M.A. Abdelkareem, T. Wilberforce, E.T. Sayed, Application of graphene in energy storage device – A review, Renew. Sustain. Energy Rev. 135 (2021) 110026. https://doi.org/10.1016/j.rser.2020.110026

[15]   R. Djara, Y. Holade, A. Merzouki, N. Masquelez, D. Cot, B. Rebiere, E. Petit, P. Huguet, C. Canaff, S. Morisset, T.W. Napporn, D. Cornu, S. Tingry, Insights from the Physicochemical and Electrochemical Screening of the Potentiality of the Chemically Synthesized Polyaniline, J. Electrochem. Soc. 167 (2020) 066503. https://doi.org/10.1149/1945-7111/ab7d40

[16]   C. Spöri, J.T.H. Kwan, A. Bonakdarpour, D.P. Wilkinson, P. Strasser, The Stability Challenges of Oxygen Evolving Catalysts: Towards a Common Fundamental Understanding and Mitigation of Catalyst Degradation, Angew. Chemie - Int. Ed. 56 (2017) 5994–6021. https://doi.org/10.1002/anie.201608601

[17]　M. Lee, J. Bae, High-Performance Fabric-based Electrochemical Capacitors Utilizing the Enhanced Electrochemistry of PEDOT : PSS Hybridized with SnO2 Nanoparticles, Bull. Korean Chem. Soc. 36 (2015) 2101–2106. https://doi.org/10.1002/bkcs.10412

[18]　M. Fang, G. Dong, R. Wei, J.C. Ho, Hierarchical nanostructures: Design for sustainable water splitting, Adv. Energy Mater. 7 (2017) 1–25. https://doi.org/10.1002/aenm.201700559

[19]　T.A. Le, Q.V. Bui, N.Q. Tran, Y. Cho, Y. Hong, Y. Kawazoe, H. Lee, Synergistic Effects of Nitrogen Doping on MXene for Enhancement of Hydrogen Evolution Reaction, ACS Sustain. Chem. Eng. 7 (2019) 16879–16888. https://doi.org/10.1021/acssuschemeng.9b04470

[20]　M. Wang, W. Zhen, B. Tian, J. Ma, G. Lu, The inhibition of hydrogen and oxygen recombination reaction by halogen atoms on over-all water splitting over Pt-TiO2 photocatalyst, Appl. Catal. B Environ. 236 (2018) 240–252. https://doi.org/10.1016/j.apcatb.2018.05.031

[21]　R. Vinodh, C. Deviprasath, C.V.V. Muralee Gopi, V.G. Raghavendra Kummara, R. Atchudan, T. Ahamad, H.-J. Kim, M. Yi, Novel 13X Zeolite/PANI electrocatalyst for hydrogen and oxygen evolution reaction, Int. J. Hydrogen Energy. 45 (2020) 28337–28349. https://doi.org/10.1016/j.ijhydene.2020.07.194

[22]　G.A. Snook, P. Kao, A.S. Best, Conducting-polymer-based supercapacitor devices and electrodes, J. Power Sources. 196 (2011) 1–12. https://doi.org/10.1016/j.jpowsour.2010.06.084

[23]　S. Anantharaj, S.R. Ede, K. Sakthikumar, K. Karthick, S. Mishra, S. Kundu, Recent Trends and Perspectives in Electrochemical Water Splitting with an Emphasis on Sulfide, Selenide, and Phosphide Catalysts of Fe, Co, and Ni: A Review, ACS Catal. 6 (2016) 8069–8097. https://doi.org/10.1021/acscatal.6b02479

[24]　A.A. Ensafi, N. Zandi-Atashbar, Z. Mohamadi, A. Abdolmaleki, B. Rezaei, Pt-Pd nanoparticles decorated sulfonated graphene-poly(3,4-ethylene dioxythiophene) nanocomposite, An efficient HER electrocatalyst, Energy. 126 (2017) 88–96. https://doi.org/10.1016/j.energy.2017.03.012

[25]　H. Shang, Z. Zhang, C. Liu, X. Zhang, S. Li, Z. Wen, S. Ji, J. Sun, MnO2@V2O5 microspheres as cathode materials for high performance aqueous rechargeable Zn-ion battery, J. Electroanal. Chem. 890 (2021) 115253. https://doi.org/10.1016/j.jelechem.2021.115253

[26]　K. Thiagarajan, D. Balaji, J. Madhavan, J. Theerthagiri, S.J. Lee, K.-Y. Kwon, M.Y. Choi, Cost-Effective Synthesis of Efficient CoWO4/Ni Nanocomposite Electrode Material for Supercapacitor Applications, Nanomaterials. 10 (2020) 2195. https://doi.org/10.3390/nano10112195

[27]　J. Theerthagiri, R.A. Senthil, P. Nithyadharseni, S.J. Lee, G. Durai, P. Kuppusami, J. Madhavan, M.Y. Choi, Recent progress and emerging challenges of transition metal sulfides based composite electrodes for electrochemical supercapacitive energy storage, Ceram. Int. 46 (2020) 14317–14345. https://doi.org/10.1016/j.ceramint.2020.02.270

[28]　X. Lu, H. Dou, B. Gao, C. Yuan, S. Yang, L. Hao, L. Shen, X. Zhang, A flexible graphene/multiwalled carbon nanotube film as a high performance electrode material for

supercapacitors, Electrochim. Acta. 56 (2011) 5115–5121.
https://doi.org/10.1016/j.electacta.2011.03.066

[29] Z.-J. Lu, S.-J. Bao, Y.-T. Gou, C.-J. Cai, C.-C. Ji, M.-W. Xu, J. Song, R. Wang, Nitrogen-doped reduced-graphene oxide as an efficient metal-free electrocatalyst for oxygen reduction in fuel cells, RSC Adv. 3 (2013) 3990. https://doi.org/10.1039/c3ra22161j

[30] M.Q. Zhao, C.E. Ren, Z. Ling, M.R. Lukatskaya, C. Zhang, K.L. Van Aken, M.W. Barsoum, Y. Gogotsi, Flexible MXene/carbon nanotube composite paper with high volumetric capacitance, Adv. Mater. 27 (2015) 339–345.
https://doi.org/10.1002/adma.201404140

[31] W. Raza, F. Ali, N. Raza, Y. Luo, K. Kim, J. Yang, Recent Advancements in Supercapacitor Technology Nano Energy Recent advancements in supercapacitor technology, Nano Energy. 52 (2018) 441–473. https://doi.org/10.1016/j.nanoen.2018.08.013

[32] K.O. Otun, M.S. Xaba, S. Zong, X. Liu, D. Hildebrandt, S.M. El-Bahy, Z.M. El-Bahy, Double linker MOF-derived NiO and NiO/Ni supercapacitor electrodes for enhanced energy storage, Colloids Surfaces A Physicochem. Eng. Asp. 634 (2022) 128019.
https://doi.org/10.1016/j.colsurfa.2021.128019

[33] J. Yang, C. Zheng, P. Xiong, Y. Li, M. Wei, Zn-doped Ni-MOF material with a high supercapacitive performance, J. Mater. Chem. A. 2 (2014) 19005–19010.
https://doi.org/10.1039/c4ta04346d

[34] S. Krishnan, A.K. Gupta, M.K. Singh, N. Guha, D.K. Rai, Nitrogen-rich Cu-MOF decorated on reduced graphene oxide nanosheets for hybrid supercapacitor applications with enhanced cycling stability, Chem. Eng. J. 435 (2022) 135042.
https://doi.org/10.1016/j.cej.2022.135042

[35] J. Banerjee, K. Dutta, M.A. Kader, S.K. Nayak, An overview on the recent developments in polyaniline-based supercapacitors, Polym. Adv. Technol. 30 (2019) 1902–1921.
https://doi.org/10.1002/pat.4624

[36] R. Bolagam, R. Boddula, P. Srinivasan, Hybrid Material of PANI with TiO 2 -SnO 2 : Pseudocapacitor Electrode for Higher Performance Supercapacitors, ChemistrySelect. 2 (2017) 65–73. https://doi.org/10.1002/slct.201601421

[37] E. Yoo, H. Zhou, Li−Air Rechargeable Battery Based on Metal-free Graphene Nanosheet Catalysts, Am. Chem. Soc. 5 (2011) 3020–3026. https://doi.org/10.1021/nn200084u

[38] A. Dutta, S. Mitra, M. Basak, T. Banerjee, A comprehensive review on batteries and supercapacitors: Development and challenges since their inception, Energy Storage. 5 (2023).
https://doi.org/10.1002/est2.339

[39] M. Zhao, B.-Q. Li, X.-Q. Zhang, J.-Q. Huang, Q. Zhang, A Perspective toward Practical Lithium–Sulfur Batteries, ACS Cent. Sci. 6 (2020) 1095–1104.
https://doi.org/10.1021/acscentsci.0c00449

[40] Y. Li, M. Chen, B. Liu, Y. Zhang, X. Liang, X. Xia, Heteroatom Doping: An Effective Way to Boost Sodium Ion Storage, Adv. Energy Mater. 10 (2020) 1–36.
https://doi.org/10.1002/aenm.202000927

[41]   Z.S. Iro, C. Subramani, S.S. Dash, A brief review on electrode materials for supercapacitor, Int. J. Electrochem. Sci. 11 (2016) 10628–10643. https://doi.org/10.20964/2016.12.50

[42]   X. Li, H. Xie, Y. Feng, Y. Qu, L. Zhai, H. Sun, X. Liu, C. Hou, All pseudocapacitive MXene-PPy//MnO2 flexible asymmetric supercapacitor, J. Mater. Sci. Mater. Electron. 34 (2023) 1878. https://doi.org/10.1007/s10854-023-11341-6

[43]   T.E. Balaji, H. Tanaya Das, T. Maiyalagan, Recent Trends in Bimetallic Oxides and Their Composites as Electrode Materials for Supercapacitor Applications, ChemElectroChem. 8 (2021) 1723–1746. https://doi.org/10.1002/celc.202100098

[44]   N.S. Shaikh, S.B. Ubale, V.J. Mane, J.S. Shaikh, V.C. Lokhande, S. Praserthdam, C.D. Lokhande, P. Kanjanaboos, Novel electrodes for supercapacitor: Conducting polymers, metal oxides, chalcogenides, carbides, nitrides, MXenes, and their composites with graphene, J. Alloys Compd. 893 (2022) 161998. https://doi.org/10.1016/j.jallcom.2021.161998

[45]   Y. Yan, T. Wang, X. Li, H. Pang, H. Xue, Noble metal-based materials in high-performance supercapacitors, Inorg. Chem. Front. 4 (2017) 33–51. https://doi.org/10.1039/c6qi00199h

[46]   Y. Wang, Y. Ding, X. Guo, G. Yu, Conductive polymers for stretchable supercapacitors, Nano Res. 12 (2019) 1978–1987. https://doi.org/10.1007/s12274-019-2296-9

Electrocatalysts and Advanced Materials for Sustainable Energy Storage   Materials Research Forum LLC
Materials Research Foundations 182 (2025) 12-26   https://doi.org/10.21741/9781644903797-2

Chapter 2

Architecture of MOF Composites for Efficient Supercapacitors

J. MANOJ[1,4,a], RAMAKRISHNAN Vishnuraj[1,2,b], MURALI Rangarajan[1,2,c],
R. SIVASUBRAMANIAN[3,d*]

[1]Center of Excellence in Advanced Materials and Green Technologies, Amrita School of
Engineering, Coimbatore, Amrita Vishwa Vidyapeetham, India

[2]Department of Chemical Engineering and Materials Science, Amrita School of Engineering
Coimbatore, Amrita Vishwa Vidyapeetham, India

[3]Department of Chemistry, School of Engineering, Amrita Vishwa Vidyapeetham, Amaravati
campus, Andhra Pradesh

[4]Department of Chemistry, Amrita School of Physical Sciences, Coimbatore, Amrita Vishwa
Vidyapeetham, India

[a]j_manoj@cb.students.amrita.edu, [b]r_vishnuraj@cb.amrita.edu, [c]r_murali@cb.amrita.edu,
[d]s_subramanian@av.amrita.edu

Abstract

Supercapacitors are considered a potential electrochemical energy storage technology possessing high power density, stability, good rate capability, and long cycle life. The applications of supercapacitors include UPS, flash cameras, and as a supplement to batteries in electric vehicles. Based on the charge storage mechanism, supercapacitors are classified into electrical double layer, pseudocapacitors, and hybrid capacitors (including battery type), respectively. Materials such as graphene, carbon nanotubes, activated carbon, etc., store a charge through a double layer formation, and on the other hand, metal oxides, conducting polymers, etc., exhibit pseudocapacitance behaviour. In this context, metal organic frameworks (MOF) are important due to their unique molecular structure and hybrid energy storage mechanism. This chapter deals with the evolution of MOFs in supercapacitor applications and the importance of different architectures in terms of structure for energy storage applications. The performance of various metal based organic frameworks for energy storage are discussed. The mechanism of charge storage using experimental and computational tools are highlighted. Finally, the challenges and prospects of MOF in supercapacitor applications are emphasized.

Keywords

Metal Organic Frameworks (MOF), Supercapacitors, Nanomaterials, Electrodes, Energy Storage

Contents

Introduction

Energy storage systems (ESS) hold a notable position in today's world for sustainable and efficient energy utilization. ESS are indispensable in the field of energy and its utilization. Depending upon the application, they were divided into 4 categories such as low-power, medium-power, network connection, and power quality [1]. The first two are small scale systems, such as flywheels (kinetic energy), fuel cells, batteries, and supercapacitors (chemical energy) that fall under these categories, whereas the third and fourth are large-scale system examples, hydraulic systems, flow batteries, and natural gas storage. Among them, supercapacitors are a class of energy storage devices with high power density and less energy density. Supercapacitors exhibit quick charge-discharge and can run for more than one lakh charge-discharge cycles. The concept of the electrical double layer laid a foundation stone for the development of electrical double layer capacitance (EDLC) in the 18th century [2]. Compared to other electrochemical energy storage systems, supercapacitors are characterized by high power density, high capacitance, use at a large range of temperatures, and long-term durability. Some of the notable applications of supercapacitors are in electric vehicles, power sources such as excavators, Radar, GPS, and memory devices, which use LED flash units [3].

Energy Storage Mechanism in Supercapacitors

Based on the charge storage mechanism, supercapacitors are broadly classified into three types: electrical double layer capacitors (EDLC), pseudocapacitors, and hybrid capacitors [4]. EDLC based supercapacitors store charge through electrostatic attraction/adsorption of ions over the electrode surface. The electrical double layer theory forms the fundamental backbone of energy storage for EDLC based capacitors. Furthermore, in pseudocapacitors, charge is stored through a faradaic process via electron transfer at the electrode-electrolyte interface. Hence, electrodes with high conductivity and fast, reversible oxidation states can act as suitable pseudocapacitors. The performance of pseudocapacitors lies between that of batteries and EDLCs as they have higher energy density and capacitance than EDLCs. Further, hybrid capacitors, on the other hand, combine the properties of EDLC and pseudocapacitance, respectively. Herein, one electrode will be an EDLC-based electrode, and the counter electrode will be a pseudocapacitive/battery-type electrode, respectively. The performance of such a capacitor will have a synergistic effect on both materials, leading to an increase in energy density, power density, capacitance, and stability of the device. Based on the type of material, hybrid supercapacitors are classified into three types: composites, asymmetric, and supercapattery.

Materials for Supercapacitors

The electrochemical performance of supercapacitors largely depends on the electrode material employed. As mentioned above, typically, three types of materials are used EDLC, pseudocapacitive, and battery type electrodes. All carbon-based electrodes, such as activated carbon, graphene, multiwalled carbon nanotubes, biomass derived carbon, hard carbon, etc. show EDLC based supercapacitors. Pseudocapacitance is exhibited by metal oxides, conducting polymers, perovskites, and spinel structures [4]. Notably, the author has worked extensively in developing biomass derived activated carbon and spinel structures such as $MnCo_2O_4$ [5], $NiCo_2O_4$ [6] for supercapacitor application. Interestingly, it was proposed that the incorporation of secondary structures such as NiO within $NiCo_2O_4$ shows enhanced performance compared to pristine $NiCo_2O_4$ electrodes. Further battery type electrodes, such as metal hydroxides/metal sulphides and other 2D materials, have also been investigated in the past. The focus is to obtain the desired electrode material with a large specific surface area, good pore size, electrical conductivity, high corrosion resistance, and facile synthesis process. Among the electrode materials MOFs are a class of compounds with high porosity and atomic level dispersion of metal centres, extensive surface area with tuneable redox centres. Recently, MOF has been explored in relation to supercapacitor applications [7]. Metals such as Co, Ni, Cu, V, Fe, etc., based on MOF have been studied recently. Nevertheless, MOF also faces challenges such as poor conductivity and electrode degradation during cycling. To overcome this challenge, MOF are also combined with other support such as carbon, graphene, conducting polymers. On the other hand, they are also used as sacrificial templates in preparing porous carbon structures. Further down the line, the importance of MOF in energy storage and studies carried out on various MOF architectures in supercapacitor application will be discussed.

Importance of Metal Organic Framework

The properties of the MOFs can be alternated by changing the metal ions or ligands, which are made up of organic acids and nitrogen containing compounds. The processes of gating and kinetic trapping are unique to flexible MOFs, which are important for adsorption purposes [8]. MOFs are prepared using various synthesis methods such as sonochemical, electrochemical, hydrothermal, mechanochemical, template-assisted, spray drying, solvothermal, etc. [9]. Typically, the synthesis of MOF is robust and requires a proper choice of metal and ligands for targeted applications. MOFs are generally employed in gas storage and separation due to their increased surface area, which promotes interaction between the metal ions in the MOFs and H_2 molecules. MOF-177 exhibits gravimetric H_2 uptake of 7.5 weight % at 70 bar and 77k [10]. In heterogeneous catalysis, for example, cross coupling of C-H compounds with electron-deficient compounds like alkene or azodicarboxylate using Zr-OTf-EY (EY-EOSIN Y DYE) [11]. Further, in LED technology, MOFs ensure superior efficiency, prolong durability, and reduced energy usage. Other notable areas, such as food packaging, photo-degradation, and anti-microbial resistance, show promising applications for MOFs [12].

MOF Architectures

Generally, MOFs are considered as modular hybrid materials wherein the crystal structure and morphology can be varied to achieve a wide range of physical and chemical properties. MOFs are considered wide-bandgap insulators; however, recently, many electrochemically active redox MOFs have been developed. $Ni_3(HITP)_2$ is the initial electrochemically active MOF reported in 2017 [13] and employed in supercapacitor applications. To date, many such redox active MOFs have been reported for energy storage applications. Based on the spatial arrangements, MOFs are classified into 0D, 1D, 2D, and 3D structures, respectively (Figure 1) [14]. The architecture of MOF plays a vital role as it dictates the pore structure as well as the surface area, which are essential prerequisites for supercapacitors. 0D MOFs can enhance material utilization and display new properties compared to their bulk counterparts. 0D MOFs can be made using techniques like rapid nucleation and limiting crystal growth in all three axes. A 0D MOF HKUST-1 made of copper and benzene-1,3,5-tricarboxylate, which exhibited 0D and Pristine MOF, acts as catalysts for enhanced organic transformation like methanolysis of styrene oxide [15].

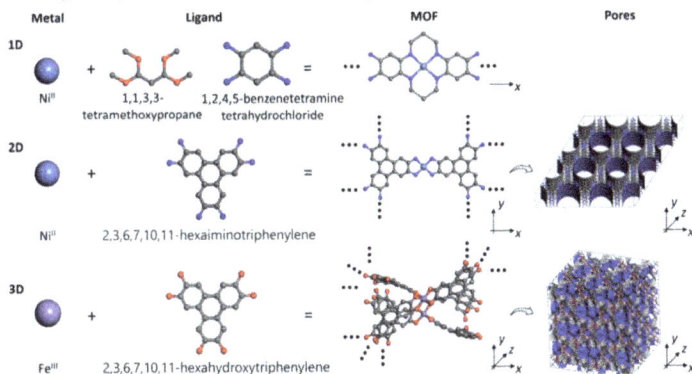

Figure 1. Dimensions of MOF (Reproduced with permission from Ref. [14] Copyright 2025, Wiley).

1D MOFs are prepared by connecting metal with bidentate ligands, which favours unidirectional growth, for example, Ni(II) with benzene triamine ligand [16]. Although intrinsic pores are difficult to achieve, the material will yield mesopores. Similarly, 2D MOFs with good electrical conductivity are prepared using π-π stacking. Such stacking develops in-plane conductivity and uniform accessible pore size. In addition, they also possess properties such as larger surface area, optical transparency, high surface to volume ratio, and atomic ultrathin thickness. On the other hand, 3D MOFs built using interconnected structures with spatial dimensions. For example, MOF-5 3D structure is prepared by the interaction of Zn with 1,4-benzodicarboxylate [17]. They are flexible, dynamic and are widely prepared and used in several applications, including supercapacitors.

MOF based Supercapacitors

As mentioned earlier, MOF based supercapacitors have recently been widely explored in various architectural and structural morphologies. The first reported MOF was based on Ni coupled with HTTP ligand in 2017, and the cell discharged from 1 V at 0.1 A/g showed a capacitance of 102 F/g. The cell was cycled 10,000 times at 1V. Similarly, MOF based on metals such as Cu, Co, Fe, Zn, Ag, etc., has been studied for supercapacitor application. Initially, monometallic based MOFs were studied. Soumen et al. [18] synthesized MOF based on succinic acid with Co (II) ion as the central metal atom. An asymmetric supercapacitor was fabricated based on Co MOF as a cathode and PPD-rGO as an anode, respectively. From the GCD studies, the specific capacitance of ASC was found to be 72.4 F/g at 2 A/g. Similarly, S.S. Shalini and A.C. Bose [19] synthesized silver-trimesic acid MOF through the refluxing method. The SEM images showed that Ag-BTC MOF has a diamond-like structure. Interestingly, it was found that during charging Ag^+ oxidizes to Ag^{2+} (unstable state), and during discharging, Ag^{2+} gains an electron and transforms to Ag^+(stable state) again. The device exhibited a specific capacitance of 408 C/g at 1 A/g with 97% capacity retention. Further, Ghosh et al. [20], have prepared three monometallic MOF from lanthanide elements Ce, Pr, and Nd, prepared by the solvothermal method with 3,3',3"-((1,3,5-triazine-2,4,6- triyl) tris(azanediyl) tribenzoic acid (H_3TATMB). Herein, compared to Pr and Nd, Ce exhibited higher specific capacitance of 572 F/g. The electrode was also tested in two electrode configurations with a capacitance of 88 F/g at a current density of 2 A/g. In another study, based on Ce and pyridine-2,4,6-tricarboxylic acid as a linker, the electrode displayed a specific capacitance of 230 F/g, and the device was stable up to 5000 charge discharge cycles [21]. Similarly, Ni-Biphenyl dicarboxylic acid (BPDC) nanoplate morphology was achieved through a hydrothermal process prepared at 4 different temperatures. It was found that the MOF exhibited a nanoplate structure; as the temperature increased, the size of the plates also increased and became sharper and edged. However, no significant change in Ni-BPDC MOF was noticed. The device exhibited a high specific capacitance of 488 F/g for Ni-BPDC prepared at 180°C [22].

Like monometallic MOF, reports on bimetallic MOF by coupling two different metals have also been reported. The presence of two metals provides more redox active sites, thereby enhancing the capacitance of the electrode. For instance, M.R. Tamtam et al. [23] have synthesized Co-Cu-MOF by the solvothermal method. Here, 2-methylimidazole was used as linker, and an asymmetric capacitor was fabricated with AC as a negative electrode. Interestingly, the device was tested in coin cells, and the bimetallic MOF showed a capacitance of 383.36 F/g higher than its monometallic counterparts. The EIS spectra before and after cycling and a plot of power density vs energy density are displayed in Figure 2 (a-b). Similarly, NiMo-MOF [24] using terephthalic acid as an organic linker showed a specific capacitance of 3811 F/g at 1A/g, which was higher than that of Ni and Mo MOF, respectively. The DFT studies carried out on NiMo-MOF suggested that Mo atoms play a prominent role in altering the electronic properties of NiMo-MOF because PDOS peak of Mo was in the Fermi level, whereas for Ni it was in the VBM. In addition, MOF built with Ni and Co showed a capacitance of 248 F/g, and the device was able to power 26 LED bulbs connected in series.

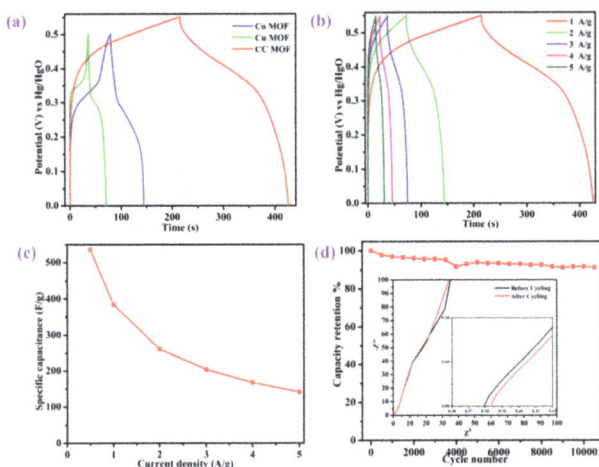

Figure 2. (a-c) GCD profile of MOF and (d) stability up to 10000 cycles and EIS spectra of CC-MOF (Reproduced with permission from Ref. [23] copyright 2025, Elsevier).

Moreover, trimetallic MOFs have gained significant attention in recent years. However, the synthesis strategy is trivial and requires a meticulous strategy for the preparation of MOF. Fatima et al. [25] prepared trimetallic composite using Co, Cu, and Ni metal ions with terephthalic acid as the organic linker. The synthesis strategy was devised in such a way that the initial trimetallic composite was treated with ammonium molybdate to yield CuCoNi-MOF@MoO₃. The incorporation of the MoO_3 resulted in the enhancement of the surface area and good thermal stability. The electrode demonstrated a specific capacitance of 218 F/g. Interestingly, hierarchically 3D-on-2D structured Ni–Co–Mn-based trimetallic MOF using 1,4 benzene dicarboxylate acid as the organic ligand was reported. MOF was coated on the Ni foam, obtained at different growth times, and labelled correspondingly. The MOF showed a maximum capacity of 1311.4 μAh cm^{-2} at a current density of 5 mA cm^{-2}. The ASC was constructed using the N and O enriched activated carbon (N–O AC) as a negative electrode. This displayed maximum power and energy density and was further utilized to harvest solar power energy for self-powered electronic applications. Rakhee Bhosale et al. [26] prepared binary-trimetallic MOFs with metal ions of Ni, Co, and Zn with two ligands, namely BDC and BTC, by the reflux condensation method. The BDC and BTC-MOF were differentiated by the SEM, and the BTC-MOF exhibited nano bricks like morphology, whereas the BDC-MOF exhibited nanoplates with irregular arrangements. Nanostructures have a greater influence on the BET adsorption and surface area, and the SSC constructed by using the BDC-MOFs exhibits a specific capacitance of 92.22 F/g with low capacitance decay and higher power and energy density.

Another interesting strategy to enhance the performance of MOF is through doping. It seems that the doped metal ions enhance the charge transfer ability and occupy the interstitial sites of MOF. A Cu doped Fe-MOF using 2-amino terephthalic acid [27] was reported to show enhanced capacitance of 562.01 F/g compared to Fe-MOF of 261 F/g, respectively. Similarly, A.A. Mahmud

17

et al. have synthesized a Cu doped Sr-MOF by BTC as an organic ligand. Doping with Cu can improve the MOFs electrical conductivity. By studying the surface morphology, it is revealed that Sr-MOF has an irregular structure. When Cu was introduced into the system, there was a brush-like growth on the surface of the Sr-MOF. Doping also results in an increased surface area of a MOF, which has a direct consequence on charge and discharge. Cu doped Sr-MOF has a specific capacitance of 243.6 C/g at 1 A/g. The GCD of Sr and Cu doped Sr MOF, Nyquist plots, and long-term cycling stability are shown in Figure 3 [28].

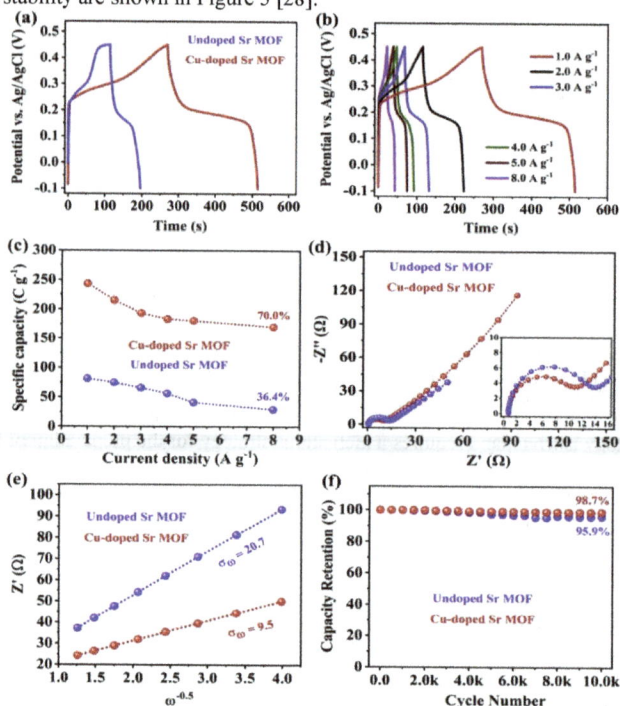

Figure 3. (a-b) GCD profile of doped and undoped MOF,(c) specific capacity of Sr and Cu@Sr, (d) Nyquist plots, (e) The relationship curves between Z' and $\omega^{-0.5}$ and (f) Long-term cycling stability at 3 A/g for Sr MOF and Cu-doped Sr MOF electrodes MOF (Reproduced with permission from Ref. [28] copyright 2025, Elsevier).

Vibhav Shukla et al. have studied the Fe doped Zn-MOF by treating the freshly synthesized Zn-MOF with $FeSO_4$ with three different molar ratios. Among them, Fe@Zn-MOF-2 exhibited a specific capacitance of 318.09 F/g, whereas the Zn MOFs displayed a lower specific capacitance of 33 F/g. The doping of Fe resulted in the facilitation of redox activity and Faradaic process [29] and S.S. Patil et al. have done the incorporation of Ag into Nd-Succinate MOF. The incorporation of Ag into Nd-Succinate MOF was done by dispersing the Nd MOF in the silver ion solution, followed by heating, stirring, and drying. XRD confirmed the incorporation of Ag-Nd Succinate

MOFs. Ag and Nd undergo oxidation in the charging process, thereby forming Ag^+ and Nd^{4+}; however, both states are unstable and undergo a reduction in the discharge process, respectively, in Ag and Nd^{3+}. The Nd MOF showed a specific capacitance of 978 F/g at 1.5 A/g, but when silver was introduced into the Nd MOFs, specific capacitance increased to 1389 F/g at 1.25 A/g [30]. It is noteworthy to mention that Rajkumar et al. [31] in 2024 reported Ni doped CaMo-MOF using benzene tricarboxylic acid as linker. By changing the concentration of Ca and Mo, 3D flower-like morphology was obtained, enabling easy intercalation/deintercalation between the layers and the diffusion process. NCMF0.5 also exhibited higher specific capacitance of 822 F/g and higher charging and discharging rate. The corresponding ASC attained energy and power density of 45 Wh/kg and 750 W/kg at 1 A/g current density.

Moreover, in recent days, transition metal dichalcogenides (TMDs) have been used in energy storage applications due to their attractive physicochemical properties. Materials like $MoTe_2$, CuS, WSe_2, etc., are the most commonly used. They have properties like higher conductivity, unique structure, and chemical stability. Thus, combining these types of materials with MOFs can increase the supercapacitor performance of the MOFs. H.B. Albargi et al. synthesized nanohybrids of Fe-MOF@$MoTe_2$ composite by solvothermal treatment. The capacitance value for Fe-MOF, $MoTe_2$, and Fe-MOF@$MoTe_2$ was estimated to be 683 C/g, 984 C/g, and 1392 C/g, respectively. These values justified the combination of the MOFs and TMDs to improve the performance of the supercapacitors. The supercapacitor device was constructed using Fe-MOF@$MoTe_2$ as a positive electrode and AC as a negative electrode (Fe-MOF@$MoTe_2$//AC). The composite displayed a 300 C/g at 1.0 A/g. A b-value for Fe-MOF@$MoTe_2$//AC is obtained between 0.5 and 0.8, which exhibits the supercapattery setup, and charge storage occurs due to the faradaic and capacitive processes. Fe-MOF@$MoTe_2$//AC also displayed a good energy and power density of 80.3 Wh/kg and 800 W/kg [32]. Similar to the (TMDs), the various monometallic and bimetallic oxides (perovskite and spinel structure) were used as the electrode materials for energy storage applications. A.U. Rehman et al. have made a spinel mixed metal oxide $CoLa_2O_4$ and a bimetallic MOF with V and Ag as metal ions and Phosphotungstic Acid as an organic ligand. The composite $CoLa_2O_4$/V-Ag-MOF was prepared by just the physical blending of MOFs and metal oxides. The Specific capacitance calculated from the GCD of V-Ag-MOF, $CoLa_2O_4$, and $CoLa_2O_4$/V-Ag-MOF was found to be 262 C/g, 786 C/g, and 1146 C/g at 2 A/g. This increase in the specific capacitance gives good evidence for the MOF and metal oxide composites. $CoLa_2O_4$/V-Ag-MOF//AC displayed a specific capacitance of 373.5 C/g at 1.1 A/g and energy density of 83.1 Wh/kg, and power density of 3360 W/kg [33]. N. Muzaffar et al. synthesized a ternary composite of $MoLa_2O_4$@CoMOF@rGO. It is known that MOFs offer an exceptional surface area despite their poor conductivity and low mechanical stability by tuning with various ligands and types of synthesis. However, a composite with a spinel can boast electrical conductivity, and rGO can offer thermal and mechanical stability to the composites. $MoLa_2O_4$@CoMOF@rGO//AC offered a capacitance of 1470 F/g at 3 mV/s. Also, the K^+ ions in electrolytes undergo an insertion and expulsion, which enhances the charging and discharging [34]. The excellent performance is mostly attributed to the synergistic effect of metal/metal oxide coupled with MOF, respectively. Table 1 gives the details of various MOF as supercapacitors.

Table 1. Illustrates the performance of various MOF based supercapacitors.

MOF	Organic linker	Electrode	Electrolyte	Specific capacitance (F/g)	Retention capacity and no of cycles	Power density (W/kg)	Energy density (Wh/kg)	Ref.
Ni-Co-Zn MOF	BDC	BDC–MOF//BDC–MOF	2M KOH	94.92 F/g at 1 mA cm^{-2}	99.2% and 3000	1367.42	47.59	[26]
Ag MOF	BTC	Ag-TA//AC	2M KOH	408 C/g at 1 A/g	97 % and 10000	650	77.46	[19]
Co-Cu MOF	BTC	CC-MOF//AC	3M KOH	203 F/g at 1 A/g	80.86% and 10000	-	55.38	[23]
Cu-Cr MOF	BDC	Cu$_{0.50}$Cr$_{0.50}$-MOF-D	2M KOH	535.1 at 0.7 A/g	94.3% and 5000	2600	36.7	[35]
Zn-Ni MOF	Glycolic acid	ZnNi-MOF(GA)//AC	1M KOH	55 at 2 A/g	86% and 10000			[36]
Me-Mn-MOF Me-Mo-MOF	Melamine	Me-Mn-MOF Me-Mo-MOF	3M KOH	Mn-MOF 653.54 Mo-MOF 312.63	Me-Mn-MOF and 10000 Me-Mo-MOF and 10000	Mn-MOF - 3048.7 Mo-MOF - 2376.6	Mn-MOF - 23.1 Mo-MOF – 12.9	[37]
Ce-BTC	BTC	CeO$_2$/rGO/CeS$_2$	3M KOH	720 at 10mV/s	3000	2917.2	23.5	[38]
Mn-BTC MOF	BTC	Mn MOF//graphene	6 mol/L KOH	89 at 0.5 A/g	91.8% and 10000			[39]
Co-BTC	BTC	Co BTC MOF/GNS	1M KOH	608.2 at 0.25 A/g	94.9% and 2000	1025.8	49.8	[40]
Mn-BDC	BDC	Mn-BDC (SSC)	1M Na$_2$SO$_4$	64.5 at 0.25 A/g	98% and 2000	171.6	4.3	[41]
Al-BDC	BDC	MIL-53 (Al) @ rGO(C-10)	1M KOH	280 at 0.5 A/g	50% and 5000	3655	6.66	[42]

Mechanism of Charge Storage

As mentioned above, the energy storage in supercapacitors can be primarily classified into EDLC and pseudocapacitance. MOF stores charge mostly by pseudocapacitance exhibiting rapid redox active sites and perform better than EDLC. However, MOF electrodes based on EDLC were also reported; such electrodes possess an inhomogeneous surface charge distribution and more complicated polarization. The charge storage mechanism in MOF has been studied extensively by experimental and theoretical approaches. For example, A.U Rehman [33] studied the ternary composite $CoLa_2O_4$/V-Ag- MOF, and found that the incorporation of OH^- ions and possible formation of metal oxyhydroxides facilitates the charge storage efficiently. The mechanism is proposed as follows Eq. (1-3)

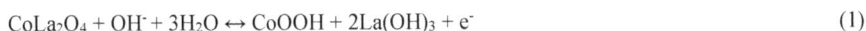

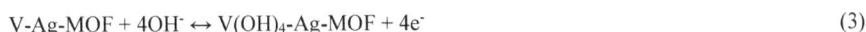

$$CoLa_2O_4 + OH^- + 3H_2O \leftrightarrow CoOOH + 2La(OH)_3 + e^- \tag{1}$$

$$CoOOH + OH^- \leftrightarrow CoO_2 + H_2O + e^- \tag{2}$$

$$\text{V-Ag-MOF} + 4OH^- \leftrightarrow V(OH)_4\text{-Ag-MOF} + 4e^- \tag{3}$$

Similarly, in an interesting report, D.B. Bailmare et al. [43] proposed the mechanism for the redox process in bimetallic MOF Ni-Co MOF, which is as follows Eq. (4-6)

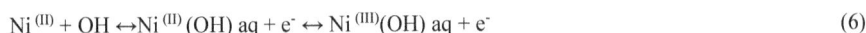

$$Co^{(II)} s + OH \leftrightarrow Co^{(II)}(OH)aq + e^- \tag{4}$$

$$Co^{(II)}(OH) aq \leftrightarrow Co^{(III)}(OH)aq + e^- \tag{5}$$

$$Ni^{(II)} + OH \leftrightarrow Ni^{(II)}(OH) aq + e^- \leftrightarrow Ni^{(III)}(OH) aq + e^- \tag{6}$$

Further, the capacitive charge and total charge contribution were determined. Based on the kinetic studies, it was reported that there was around 60% pseudocapacitance contribution from the MOF electrode. Similar reports provide insightful interest in the mechanism of MOF electrodes in supercapacitors.

Further deep insights can be realized through materials modelling, and simulation through density functional theory (DFT) about the charge transfer at the electrode-electrolyte interface. In addition, *in situ* experimental approaches such as NMR, Raman spectroscopy, small angle x-ray, small angle neutron scattering techniques and electrochemical quartz crystal microbalance (EQCM) were employed to study the charge storage mechanism of supercapacitors, especially the thermodynamic and kinetic mechanisms of electrodes [44]. Cheng et al. [45] studied the charging mechanism of pseudocapacitance MOF using EQCM coupled with IR and XPS spectroscopy. In addition, the studies with respect to the stability of MOF using degradation studies were carried out using X-ray absorption near edge structure measurements (XANES) technique. The rapid redox transition of metal at the electrode-electrolyte interface was studied using this technique. Further Molecular Dynamic Simulation studies were also carried out on 2D MOF to study the charging mechanism, wherein the polarization mechanism was upheld.

Future challenges

The performance of MOF as an efficient supercapacitor electrode has been demonstrated over the last few years. MOF electrodes with high capacitance and excellent cycling performances were reported earlier. However, due to several limitations, the performance of MOF has not been on par with standard/benchmark materials such as porous carbon YP-50F. This is because most of the MOF electrodes exhibited poor potential windows in the range of $1 - 1.5$ V compared to carbon materials (~2V). Further, the material exhibits poor kinetics or rate performance and lower cycling stability compared to other standard materials. For instance, $Ni_3(HITP)_2$ based MOF electrodes exhibit ~85 to 90 % of initial capacity after 10000 cycles. On the other hand, porous carbons exhibit more than 99 %. Despite the odds, the field opens up exciting potential in developing MOF for supercapacitor applications. The material can be improved further by fine tuning the chemistry of the metal and organic linkers so that a suitable framework/dimension can be achieved, which in turn impacts the pore size of the MOFs. A few strategies, such as the incorporation of specific functional groups to improve the electrode-electrolyte interaction, structure correlation, and increase the pseudocapacitive contribution, can aid in further enhancing the performance of MOF electrodes. In addition, with the development of *in situ* techniques, the charging and degradation mechanism can be explored. Further computational or simulation studies using DFT or MD simulations/AI tools such as machine learning and deep learning can be employed to study the ion transport and electrochemical performances of MOF based electrodes.

Conclusion

To conclude, this chapter summarizes the need for energy storage devices and the significance of supercapacitors compared to batteries. The fundamental charge storage mechanism of supercapacitors and the classification based on their storage have been discussed. The contribution of various materials, such as carbon based, conducting polymers, metal oxides, and mixed metal oxides, was noted, and the importance of MOFs in energy storage was highlighted. Various MOF architectures and strategies for fine-tuning the structures were discussed. In addition, the performance of MOF in terms of capacitance, rate capability, and stability was explored. Moreover, the charge storage mechanism of MOF pertaining to both theoretical and experimental studies, which includes *in-situ* IR, NMR techniques etc., was explained. Ultimately, the challenges and future perspectives were discussed to fully harness the potential MOFs in supercapacitor applications.

References

[1]H. Ibrahim, A. Ilinca, J. Perron, Energy storage systems-Characteristics and comparisons, Renew. Sustain. Energy Rev. 12 (2008) 1221–1250. https://doi.org/10.1016/j.rser.2007.01.023

[2]P. Bhojane, Recent advances and fundamentals of Pseudocapacitors: Materials, mechanism, and its understanding, J. Energy Storage 45 (2022) 103654. https://doi.org/10.1016/j.est.2021.103654

[3]G. Gautham Prasad, N. Shetty, S. Thakur, Rakshitha, K.B. Bommegowda, Supercapacitor technology and its applications: A review, IOP Conf. Ser. Mater. Sci. Eng. 561 (2019). https://doi.org/10.1088/1757-899X/561/1/012105

[4] S. Sharma, P. Chand, Supercapacitor and electrochemical techniques: A brief review, Results Chem. 5 (2023) 100885. https://doi.org/10.1016/j.rechem.2023.100885

[5] M. Haripriya, A.M. Ashok, S. Hussain, R. Sivasubramanian, Nanostructured MnCo2O4 as a high-performance electrode for supercapacitor application, Ionics (Kiel). 27 (2021) 325-337. https://doi.org/10.1007/s11581-020-03788-y

[6] M. Haripriya, R. Sivasubramanian, A.M. Ashok, S. Hussain, G. Amarendra, Hydrothermal synthesis of NiCo2O4-NiO nanorods for high performance supercapacitors, J. Mater. Sci. Mater. Electron. 30 (2019) 7497–7506. https://doi.org/10.1007/s10854-019-01063-z

[7] B. Chettiannan, E. Dhandapani, G. Arumugam, R. Rajendran, M. Selvaraj, Metal-organic frameworks: A comprehensive review on common approaches to enhance the energy storage capacity in supercapacitor, Coord. Chem. Rev. 518 (2024) 216048. https://doi.org/10.1016/j.ccr.2024.216048

[8] A.J. Fletcher, K.M. Thomas, M.J. Rosseinsky, Flexibility in metal-organic framework materials: Impact on sorption properties, J. Solid State Chem. 178 (2005) 2491–2510. https://doi.org/10.1016/j.jssc.2005.05.019

[9] D. Wang, H. Yao, J. Ye, Y. Gao, H. Cong, B. Yu, Metal-Organic Frameworks (MOFs): Classification, Synthesis, Modification, and Biomedical Applications, Small 2404350 (2024) 1–53. https://doi.org/10.1002/smll.202404350

[10] H. Furukawa, M.A. Miller, O.M. Yaghi, Independent verification of the saturation hydrogen uptake in MOF-177 and establishment of a benchmark for hydrogen adsorption in metal-organic frameworks, J. Mater. Chem. 17 (2007) 3197–3204. https://doi.org/10.1039/b703608f

[11] D. Li, A. Yadav, H. Zhou, K. Roy, P. Thanasekaran, C. Lee, Advances and Applications of Metal-Organic Frameworks (MOFs) in Emerging Technologies: A Comprehensive Review, Glob. Challenges 8 (2024) 1–32. https://doi.org/10.1002/gch2.202300244

[12] P. Ananthi, K. Hemkumar, A. Pius, Antibacterial, Biodegradable Polymeric Films Loaded with Co-MOF/ZnS Nanoparticles for Food Packaging and Photo-Degradation Applications, ACS Food Sci. Technol. 4 (2024) 1462–1471. https://doi.org/10.1021/acsfoodscitech.4c00087

[13] D. Sheberla, J.C. Bachman, J.S. Elias, C.J. Sun, Y. Shao-Horn, M. Dincă, Conductive MOF electrodes for stable supercapacitors with high areal capacitance, Nat. Mater. 16 (2017) 220–224. https://doi.org/10.1038/nmat4766

[14] S.J. Shin, J.W. Gittins, C.J. Balhatchet, A. Walsh, A.C. Forse, Metal–Organic Framework Supercapacitors: Challenges and Opportunities, Adv. Funct. Mater. 2308497 (2023) 1–11. https://doi.org/10.1002/adfm.202308497

[15] L.H. Wee, M.R. Lohe, N. Janssens, S. Kaskel, J.A. Martens, Fine tuning of the metal-organic framework Cu 3(BTC) 2 HKUST-1 crystal size in the 100 nm to 5 micron range, J. Mater. Chem. 22 (2012) 13742–13746. https://doi.org/10.1039/c2jm31536j

[16] C. Wang, Y.V. Kaneti, Y. Bando, J. Lin, C. Liu, J. Li, Y. Yamauchi, Metal-organic framework-derived one-dimensional porous or hollow carbon-based nanofibers for energy

storage and conversion, Mater. Horizons 5 (2018) 394–407.
https://doi.org/10.1039/c8mh00133b

[17]   H. Liu, Y. Zhao, B. Huang, H. Liu, P. Zhang, W. Gu, T. Ma, Zn-Based Three-Dimensional Metal-Organic Framework for Selective Fluorescence Detection in Zwitterionic Ions, (2025) 1–14.

[18]   S. Khan, S. Halder, S. Chand, A.K. Pradhan, C. Chakraborty, Co-containing metal-organic framework for high-performance asymmetric supercapacitors with functionalized reduced graphene oxide, Dalt. Trans. 52 (2023) 14663–14675.
https://doi.org/10.1039/d3dt02314a

[19]   S.S. Shalini, A.C. Bose, Design and development of diamond-shaped Silver-Trimesic acid based Metal-Organic framework for high-performance supercapacitor application, J. Electroanal. Chem. 951 (2023) 117895. https://doi.org/10.1016/j.jelechem.2023.117895

[20]   S. Ghosh, A. De Adhikari, J. Nath, G.C. Nayak, H.P. Nayek, Lanthanide (III) Metal-Organic Frameworks: Syntheses, Structures and Supercapacitor Application, ChemistrySelect 4 (2019) 10624–10631. https://doi.org/10.1002/slct.201902614

[21]   M. Shahbaz, S. Sharif, A. Shahzad, Z.S. Şahin, B. Riaz, S. Shahzad, Enhanced electrochemical performance of cerium-based metal organic frameworks derived from pyridine-2,4,6-tricarboxylic acid for energy storage devices, J. Energy Storage 88 (2024).
https://doi.org/10.1016/j.est.2024.111463

[22]   W. Zhang, H. Yin, Z. Yu, X. Jia, J. Liang, G. Li, Y. Li, K. Wang, Facile Synthesis of 4, 4′-biphenyl Dicarboxylic Acid-Based Nickel Metal Organic Frameworks with a Tunable Pore Size towards High-Performance Supercapacitors, Nanomaterials 12 (2022).
https://doi.org/10.3390/nano12122062

[23]   M.R. Tamtam, R. Koutavarapu, R. Wang, G.S. Choi, J. Shim, Cobalt–copper MOF: A high-performance and ecofriendly electrode material for symmetric and asymmetric supercapacitors, Mater. Sci. Semicond. Process. 188 (2025) 109220.
https://doi.org/10.1016/j.mssp.2024.109220

[24]   S. Kanthasamy, M. Subramani, S. Ramasamy, S. Thangavelu, Unveiling the structure-property relationships of bimetallic nickel molybdenum metal organic framework for pseudocapacitor electrode materials: A combined approach of experimental and theoretical study, Chem. Eng. J. 495 (2024) 153691. https://doi.org/10.1016/j.cej.2024.153691

[25]   S. Fatima, H. Shabbir, R. Sharif, H.M. Fahad, J. Yang, F. Shaheen, R. Wahab, S. Akbar, V. Perumal, A novel binary composite of CuCoNi-MOF/MoO3 with exceptional capacitance as electrode material for supercapacitors, J. Energy Storage 99 (2024).
https://doi.org/10.1016/j.est.2024.113300

[26]   R. Bhosale, S. Bhosale, D. Narale, C. Jambhale, S. Kolekar, Construction of Well-Defined Two-Dimensional Architectures of Trimetallic Metal-Organic Frameworks for High-Performance Symmetric Supercapacitors, Langmuir 39 (2023) 12075–12089.
https://doi.org/10.1021/acs.langmuir.3c01337

[27]   S.A. Patil, P.K. Katkar, M. Kaseem, G. Nazir, S.W. Lee, H. Patil, H. Kim, V.K. Magotra, H.B. Thi, H. Im, N.K. Shrestha, Cu@Fe-Redox Capacitive-Based Metal–Organic Framework

Film for a High-Performance Supercapacitor Electrode, Nanomaterials 13 (2023). https://doi.org/10.3390/nano13101587

[28] A. Al Mahmud, A.H. Alshatteri, H.S. Alhasan, W. Al Zoubi, K.M. Omer, M.R. Thalji, Copper-doped strontium metal-organic framework: Dual-function active material for supercapacitor and oxygen evolution reaction, Electrochim. Acta 503 (2024) 144857. https://doi.org/10.1016/j.electacta.2024.144857

[29] V. Shukla, N. Waris, M.Z. Khan, K.A. Siddiqui, Fabrication of an iron-doped Fe@Zn-MOF composite: empowering enhanced colorimetric recognition and energy storage performance, New J. Chem. 48 (2024) 11518–11529. https://doi.org/10.1039/d4nj01896f

[30] S.S. Patil, A.C. Khandare, B.S. Khillare, P.R. Kagne, V.N. Narwade, T. Hianik, M.D. Shirsat, Incorporating Ag into Nd-Succinate MOF: Tunned electrochemical supercapacitor behaviour of Ag modified Nd-Succinate and electrochemical reaction mechanism, Colloids Surfaces A Physicochem. Eng. Asp. 703 (2024) 1–10. https://doi.org/10.1016/j.colsurfa.2024.135215

[31] P. Rajkumar, V. Thirumal, A.S. Rasappan, M.S. Iyer, S. Asaithambi, K. Yoo, J. Kim, Electrochemically enhanced battery-type Ni substituted CaMo-MOF electrodes: Towards futuristic energy storage system, J. Energy Storage 80 (2024) 110284. https://doi.org/10.1016/j.est.2023.110284

[32] H.B. Albargi, A. Abbas, M. Zeeshan, M.W. Iqbal, N.A. Ismayilova, M.A. Sunny, H. Hassan, T. Abbas, Design of iron-based metal-organic framework (Fe-MOF) and molybdenum telluride (MoTe2) nanohybrids for enhanced energy storage and hydrogen evolution reactions, Inorg. Chem. Commun. 173 (2025) 113791. https://doi.org/10.1016/j.inoche.2024.113791

[33] A.U. Rehman, N. Muzaffar, I. Barsoum, A.M. Afzal, M. Ali, M.W. Iqbal, Z. Ahmad, S. Mumtaz, A.A.A. Bahajjaj, S.A. Munnaf, Designing of efficient CoLa2O4/V-Ag-MOF hybrid electrode for energy storage, hydrogen evolution reaction, and chemical sensors, Fuel 384 (2025) 133991. https://doi.org/10.1016/j.fuel.2024.133991

[34] N. Muzaffar, I. Barsoum, A.M. Afzal, M.W. Iqbal, Z. Ahmad, A.S. Alqarni, S.A.M. Issa, H.M.H. Zakaly, Exploring the electrochemical potential of MoLa2O4/rGO/Co-MOF nanocomposites in energy storage and monosodium glutamate detection, Fuel 385 (2025) 134131. https://doi.org/10.1016/j.fuel.2024.134131

[35] S.A. Al-Thabaiti, M.M.M. Mostafa, A.I. Ahmed, R.S. Salama, Synthesis of copper/chromium metal organic frameworks - Derivatives as an advanced electrode material for high-performance supercapacitors, Ceram. Int. 49 (2023) 5119–5129. https://doi.org/10.1016/j.ceramint.2022.10.029

[36] M.A. Sunny, H. Hassan, B.S. Almutairi, E. Umar, M.W. Iqbal, A.K. Alqorashi, H. Alrobei, N. Ahmad, N.A. Ismayilova, Synthesis of Eco-friendly Zn-Ni bimetallic MOFs with biodegradable glycolic acid ligands for enhanced supercapacitor performance and hydrogen evolution reaction, Phys. Scr. 99 (2024). https://doi.org/10.1088/1402-4896/ad74a4

[37] R. Vanaraj, B. Arumugam, G. Mayakrishnan, R. Kanthapazham, S.C. Kim, Specific Capacitance Enhancement of Metal-Organic Framework (MOF) by Boosting Intramolecular

Charge Transfer Mechanism, ACS Appl. Energy Mater. (2024).
https://doi.org/10.1021/acsaem.4c01241

[38]   U.A. Khan, N. Iqbal, T. Noor, R. Ahmad, A. Ahmad, J. Gao, Z. Amjad, A. Wahab,
Cerium based metal organic framework derived composite with reduced graphene oxide as
efficient supercapacitor electrode, J. Energy Storage 41 (2021) 102999.
https://doi.org/10.1016/j.est.2021.102999

[39]   J. Xu, Y. Yang, Y. Wang, J. Cao, Z. Chen, Enhanced electrochemical properties of
manganese-based metal organic framework materials for supercapacitors, J. Appl.
Electrochem. 49 (2019) 1091–1102. https://doi.org/10.1007/s10800-019-01352-9

[40]   N.S. Punde, C.R. Rawool, A.S. Rajpurohit, S.P. Karna, A.K. Srivastava, Hybrid
Composite Based on Porous Cobalt-Benzenetricarboxylic Acid Metal Organic Framework
and Graphene Nanosheets as High Performance Supercapacitor Electrode, ChemistrySelect 3
(2018) 11368–11380. https://doi.org/10.1002/slct.201802721

[41]   S. Sundriyal, S. Mishra, A. Deep, Study of manganese-1,4-benzenedicarboxylate metal
organic framework electrodes based solid state symmetrical supercapacitor, Energy Procedia
158 (2019) 5817–5824. https://doi.org/10.1016/j.egypro.2019.01.546

[42]   M. Majumder, R.B. Choudhary, A.K. Thakur, A. Khodayari, M. Amiri, R. Boukherroub,
S. Szunerits, Aluminum based metal-organic framework integrated with reduced graphene
oxide for improved supercapacitive performance, Electrochim. Acta 353 (2020) 136609.
https://doi.org/10.1016/j.electacta.2020.136609

[43]   D.B. Bailmare, K.N. Pande, D. Peshwe, S. Chopra, A.D. Deshmukh, Boosting capacitive
performance of electrode material by facile incorporation of phthalic acid ligand-based
bimetallic MOF for supercapacitors, Electrochim. Acta 503 (2024) 144856.
https://doi.org/10.1016/j.electacta.2024.144856

[44]   K. Xu, H. Shao, Z. Lin, C. Merlet, G. Feng, J. Zhu, P. Simon, Computational Insights into
Charge Storage Mechanisms of Supercapacitors, Energy Environ. Mater. 3 (2020) 235–246.
https://doi.org/10.1002/eem2.12124

[45]   S. Cheng, W. Gao, Z. Cao, Y. Yang, E. Xie, J. Fu, Selective Center Charge Density
Enables Conductive 2D Metal−Organic Frameworks with Exceptionally High
Pseudocapacitance and Energy Density for Energy Storage Devices, Adv. Mater. 34 (2022)
1–11. https://doi.org/10.1002/adma.202109870

Electrocatalysts and Advanced Materials for Sustainable Energy Storage   Materials Research Forum LLC
Materials Research Foundations 182 (2025) 27-38   https://doi.org/10.21741/9781644903797-3

Chapter 3

Exploring Electrocatalytic Excellence of MXene based Catalysts for Supercapacitance Studies

DICKSON D. Babu[1,a*], PRAVEEN Naik[2,b*], SUJIN Jose[3,c], SANTHOSH T.C.M.[4,d]

[1]Department of Chemistry, St. Thomas College, Kozhencherry, 689641, Kerala, India

[2]Department of Chemistry, Nitte Meenakshi Institute of Technology, Nitte (Deemed to be University), Bengaluru Campus, Bengaluru 560064, India

[3]Advanced Materials Laboratory, School of Physics, Madurai Kamaraj University, Madurai 625021, India

[4]Departement of Physics, Akash Institute of Engineering and Technology, Affiliated to Visvesvaraya Technological University, Belgavi Prasannahalli Main Road, Devanahalli - 562110

[a]dicksondbabu@stthomascollege.info, [b]praveennaik018@gmail.com, [c]sujamku@gmail.com, [d]santhu.990m@gmail.com

Abstract

Ever since their discovery in 2011, MXene-based materials have attracted attention as a novel electrocatalyst. They have outstanding electrical conductivity, adjustable surface chemistry, and versatile structures. These qualities make them ideal candidates to transform the next generation of supercapacitors. This chapter explores the role of MXenes in supercapacitor applications, focusing mainly on their basic charge storage mechanisms, which include electric double-layer capacitance (EDLC) and pseudocapacitance. These materials show fast charge transfer and ion movement by leveraging MXenes' unique physicochemical properties, such as their large surface area, many redox-active sites, and customized surface modifications. This results in excellent electrochemical performance. The chapter addresses these challenges and offers cutting-edge solutions, including new synthesis techniques and surface modifications to reduce limitations. Experimental findings and comparisons with traditional materials provide a clear view of MXenes' potential in high-performance energy storage systems. The final section also suggests possible future directions, highlighting the role of MXene-based electrocatalysts in creating next-generation supercapacitors that deliver high power and long cycling life for sustainable energy applications.

Keywords

MXene, Electrocatalysis, Supercapacitors, Energy Storage

Contents

Materials Research Foundations 182 (2025) 27-38   https://doi.org/10.21741/9781644903797-3

Introduction

The demand for energy is rising quickly due to urbanization, industrial growth, and increasing population, which has created a strong need for efficient, sustainable energy storage systems [1-3]. Fossil fuels, still widely used, raise serious environmental concerns and face regular limitations in global availability [4]. Global efforts are pushing hard for cleaner energy alternatives, gaining momentum around the world. Supercapacitors are attracting significant interest because they charge and discharge quickly, provide high power output, and have a long lifespan with a minimal environmental impact [5-6].

Supercapacitors store energy through two main mechanisms: electric double-layer capacitance and pseudocapacitance. Electric double-layer capacitance relies on ions being adsorbed at the electrode-electrolyte boundary [7-9]. Pseudo capacitance, on the other hand, comes from fast, reversible electrochemical reactions near the electrode surface. Researchers are focusing on finding new materials that offer high energy and power densities along with strong mechanical stability in electrodes. MXenes show great promise as electrode materials, which were discovered in 2011 by extracting elements from MAX phases, which have the formula Mn+1AXn. In this formula, M usually stands for a transition metal, A represents a group 13 or 14 element, and X is either carbon or nitrogen [10-12]. M typically refers to an early transition metal from groups 3 to 6. Subsequent MXene structures, labelled Mn+1XnTx due to surface groups like –O, –OH, or –F, show remarkable conductivity and adjustable mechanical strength, which benefits energy storage.

Among the different MXenes studied, Ti_3C_2Tx is the most widely researched, typically derived from Ti_3AlC_2 by laboriously etching out the aluminum content [13]. The layered structure allows ions to move freely across a high surface area, which supports efficient charge storage. However, pure MXenes have some drawbacks, such as many new materials. The layers tend to stick together

Electrocatalysts and Advanced Materials for Sustainable Energy Storage   Materials Research Forum LLC
Materials Research Foundations 182 (2025) 27-38   https://doi.org/10.21741/9781644903797-3

strongly, and energy density can be somewhat limited. Structural instability can also develop over many cycles. Since their discovery, MXenes have been thoroughly examined, leading to many breakthroughs in synthesis methods and applications [14-15]. MXenes are often combined with other materials to form composites, thereby overcoming inherent performance limitations. Mixing them with carbon-based nanomaterials like graphene or carbon nanotubes can reduce layer restacking and significantly increase surface area. These combinations improve electrical conductivity and mechanical strength by allowing ions to move freely within the material. MXene-based composites show greatly enhanced electrochemical performance and hold significant promise for use in advanced supercapacitor systems.

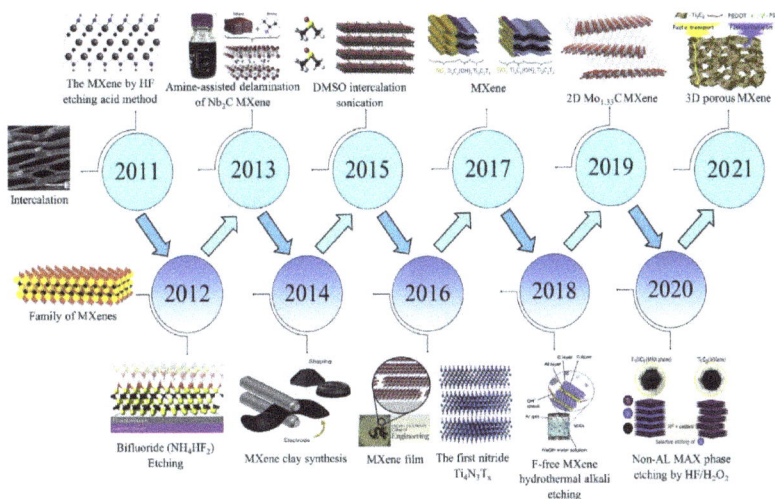

Figure 1. Timeline of MXene development from 2011 to 2021, illustrating key milestones in synthesis, structural understanding, and applications in energy storage (Reproduced with permission from Ref.[13] Copyright 2017, American Chemical Society).

MXenes frequently get paired with other materials, forming composites, thereby overcoming inherent performance limitations effectively. Mixing them with carbon-based nanomaterials like graphene or carbon nanotubes can reduce restacking of layers and noticeably increase surface area[16-17]. These combinations enhance electrical conductivity and mechanical durability quite significantly by facilitating free ion movement inside the material. MXene-based composites exhibit markedly enhanced electrochemical performance and hold considerable promise for deployment in cutting-edge supercapacitor systems nowadays. The timeline illustrating the key developments of MXenes for various applications in energy storage is given in Figure 1[15].

Recently, MXenes have gained attention for their high energy storage capabilities across various supercapacitor applications [18-19]. This includes details about their preparation methods and structural features, along with various modification techniques focused on improving overall function. Key challenges, such as material degradation and manufacturing scalability, are discussed alongside current innovations that are paving the way for new energy storage devices that are both efficient and reliable.

Electrocatalytic Properties of MXenes for Supercapacitance

MXenes exhibit remarkably strong electrocatalytic activity and possess highly adjustable surface chemistry, making them superb materials for supercapacitor devices. Pseudocapacitive activity is quite strong, owing largely to redox-active transition metal sites and surface groups like –O and –OH, enabling reversible redox reactions. Elevated metallic conductivity in them guarantees rapid electron transport, whereas the layered nature enables effective ion adsorption and intercalation, facilitating Faradaic processes and electric double-layer capacitance. Ti_3C_2Tx MXenes exhibit exceptionally high pseudocapacitance around 400 F/g in H_2SO_4, largely due to proton-coupled redox reactions happening effectively [20-22]. Further enhancements emerge from surface modification via nitrogen doping or mixing MXenes with other electrocatalytic compounds like Co_3O_4, which accelerates reaction kinetics and boosts cycling stability significantly [23-24].

The electrical conductivity of MXenes, which is around 10,000 S/cm for Ti_3C_2Tx, relies on factors like the transition metal, surface terminations, synthesis method, layer thickness, and structural defects [25-26]. Thinner flakes are more conductive but are also more prone to oxidation. Exposure to air or moisture can degrade conductivity over time. Methods such as controlled synthesis, inert storage, and defect passivation are used to address these issues.

The electrical conductivity of MXenes is generally about 10,000 S/cm for Ti_3C_2Tx, influenced mainly by the transition metal and surface terminations [25]. Thinner flakes have better conductivity but oxidize easily. Exposure to air or moisture gradually reduces conductivity. Effective solutions include defect passivation, inert storage, and controlled synthesis through various innovative methods. MXenes function well in electric double-layer capacitors (EDLC) and pseudocapacitance based on the electrolyte composition and operating conditions [27-28]. Proton adsorption and redox processes enable charge storage in acidic electrolytes like H_2SO_4, while cation intercalation occurs in neutral electrolytes such as K_2SO_4, which increases capacitance [29]. Surface functionalization is crucial in determining device performance across different operating conditions.

Functional groups added during synthesis, including oxygen, fluorine, and hydroxyl, greatly affect conductivity, wettability, and redox behavior. The presence of –O groups enhances redox activity, while –OH groups aid in proton transport. The –F groups offer some oxidation resistance but limit ion access [30-31]. Post-processing treatments like sulfidation and polymer grafting further adjust properties, and doping with nitrogen helps modify the electronic structure. MXenes can easily integrate with other materials to overcome issues like restacking and severe degradation. Hybrid frameworks that include graphene, carbon nanotubes, or porous carbon largely prevent layer restacking and create highly conductive 3D networks that improve performance [32]. Combining metal oxides such as MnO_2 and RuO_2 significantly enhances Faradaic contributions. MXene/MnO_2 composites achieve impressive specific capacitances of about 1200 F/g [33]. Conductive polymer composites with polyaniline and PEDOT: PSS greatly improve flexibility and proton conductivity, making them suitable for wearable devices [34]. MXenes are positioning themselves as a leading material for future supercapacitors, demonstrating stable performance over thousands of cycles with notably high energy densities.

Mechanisms of Electrocatalysis in MXene-based Supercapacitors

Electrode-electrolyte interface and charge transfer dynamics

MXene-based supercapacitors benefit greatly from their unique structure and chemistry at the electrode-electrolyte interfaces. MXenes have a 2D layered structure and hydrophilic surface terminations like –O, –OH, and –F. These features promote efficient ion adsorption and surface redox reactions [11,35]. Functional groups support two key mechanisms: electric double-layer capacitance and pseudocapacitance, which improve overall charge storage capacity. MXenes also have very high electrical conductivity, allowing for quick electron transfer while accommodating various electrolyte ions like H^+ in acidic environments or K^+ in neutral organic systems [36]. Oxygen-terminated surfaces assist Ti^{3+}/Ti^{4+} redox transitions, which greatly enhance pseudocapacitive behavior in these systems. However, excessive F terminations can hinder ion mobility under certain conditions.

MXene sheet restacking often restricts ion accessibility. Introducing spacers like carbon nanotubes or graphene can help address this problem. Defects in the MXene structure add redox-active sites, but it is crucial to manage them carefully to avoid disrupting electron pathways entirely. The composition and pH of the electrolyte significantly influence behavior at the interface by changing surface group protonation and affecting ion kinetics. Advanced hybridization strategies and surface functionalization methods can greatly reduce resistance, improving capacitance to nearly 400 F/g and power density significantly over 10000 cycles. The main factors influenced by the electrode-electrolyte interface include:

(i) Ion adsorption/desorption kinetics, which directly affect charge storage efficiency.

(ii) Electrochemical stability, which affects long-term durability.

(iii) Charge transfer resistance, which impacts power output and energy density.

(iv) Capacitance modulation, through interface engineering and electrolyte adjustments.

Faradaic and non-Faradaic processes in supercapacitance

MXene-based supercapacitors incorporate both non-Faradaic (EDLC) and Faradaic (pseudocapacitive) mechanisms, leading to higher energy and power densities. The EDLC process relies on the electrostatic adsorption of ions such as H^+ and K^+ onto the conductive surfaces of MXenes. Their large surface area, hydrophilicity, and accessible interlayer spacing facilitate this process. It supports quick charge-discharge cycles with excellent longevity. In comparison, pseudocapacitance includes reversible Faradaic reactions, such as the $Ti^{3+} \leftrightarrow Ti^{4+}$ transition in $Ti_3C_2T_x$, which adds to higher energy density [37-38]. These reactions occur more readily in acidic or proton-rich environments (like H_2SO_4), where protons are involved in surface redox chemistry. Neutral and alkaline electrolytes favor intercalation-based storage (such as Na^+, K^+), which contributes less to pseudocapacitance. Hybrid systems—like $MXene/MnO_2$ composites—enhance Faradaic behavior by adding more redox-active materials [39]. While MXene sheet restacking can slow down ion transport, structural engineering (such as 3D designs) boosts ion diffusion and balances EDLC and Faradaic contributions. This dual functionality allows MXenes to surpass traditional EDLC materials like activated carbon (~200 F/g), achieving ~400 F/g while maintaining competitive power and energy densities [40]. Understanding the differences between

Faradaic and non-Faradaic mechanisms is crucial for optimizing materials. The key differences are summarized in Table 1.

Table 1. Comparison of Faradaic and Non-Faradaic Charge Storage in MXenes

Feature	EDLC (Non-Faradaic)	Pseudocapacitance (Faradaic)
Charge Storage	Electrostatic ion adsorption	Surface redox reactions
Charge Transfer	No charge transfer (physical)	Involves charge transfer (chemical)
Capacitance	Moderate	High, due to redox activity
Power Density	High (fast kinetics)	Moderate
Energy Density	Low	Higher
Cycle Life	Excellent (>100,000 cycles)	Moderate (~10,000–50,000 cycles)
Voltage Window	Limited	Broader with tailored electrolytes

Role of MXenes in facilitating faster electron and ion transport

The efficient charge transport properties of MXenes arise from their unique layered structure, metallic conductivity, and functionalized surface chemistry. Their electrical and ionic transport mechanisms can be explained by the following factors:

Electron Transport in MXenes

(i) Metallic conductivity: MXenes (e.g., $Ti_3C_2T_x$) exhibit high electrical conductivity (~10,000 S/cm), which is comparable to metals.

(ii) 2D layered structure: Electrons can move freely within the delaminated MXene sheets, reducing resistance.

(iii) Charge delocalization: The presence of transition metal d-orbitals facilitates efficient electron mobility.

Ion Transport in MXenes

(ii) Interlayer spacing: The adjustable interlayer distance allows for efficient ion intercalation and deintercalation, reducing diffusion resistance.

(ii) Surface functional groups: (-OH, -F, -O): These hydrophilic groups enhance ion accessibility and interaction with electrolytes.

(iii) Hydrophilicity: MXenes possess excellent wettability, ensuring better ion transport in aqueous and organic electrolytes.

Comparative Study: MXene-based Catalysts vs. Conventional Catalysts

MXenes are emerging as competitive electrocatalysts for applications in energy conversion, environmental remediation, and chemical synthesis [35]. Conventional catalysts—including noble metals (Pt, Pd), transition metal oxides, and carbon-based materials—are effective but suffer from limitations such as high cost, limited abundance, and susceptibility to degradation. In contrast, MXenes offer a favorable combination of high electrical conductivity, large surface area, and chemically tunable active sites. Table 2 compares the critical properties of MXene-based and conventional catalysts. MXenes thus offer a cost-effective, high-performance alternative to conventional catalysts, with the potential for broad integration into next-generation energy and catalytic systems.

Table 2. Comparison of Key Properties of MXene-Based and Conventional Catalysts

Property	MXene-Based Catalysts	Conventional Catalysts
Composition	Transition metal carbides, nitrides, and carbonitrides	Noble metals (Pt, Pd), transition metal oxides, carbon materials
Electrical Conductivity	High (~10,000 S/cm for $Ti_3C_2T_x$)	Moderate to high (varies by material)
Surface Area	Large, tunable by interlayer engineering	High in nanoparticle or porous forms
Active Sites	Exposed transition metal atoms and surface functional groups (-OH, -F, -O)	Metal atoms, defect sites, and oxygen vacancies
Stability	Chemically stable, but oxidation-sensitive	Varies; some suffer from deactivation or sintering
Cost	Relatively lower than noble metal catalysts	Expensive (e.g., Pt, Pd)

Challenges in MXene-Based Supercapacitors

Despite their excellent electrical conductivity, high surface area, and versatile surface chemistry, MXene materials face several hurdles that limit their commercial use in supercapacitors [6,41,42].

Material Stability Challenges: MXenes break down quickly when exposed to air and easily oxidize in the presence of moisture, which greatly reduces their electrical conductivity and performance over time. The tendency to oxidize and poor stability during charge-discharge cycles limit the long-term use of these materials. Layer restacking and mechanical delamination often reduce surface area and slow down ion transport during electrochemical cycling. Modifying the surface with organic molecules or metal oxides significantly improves oxidation resistance. Using cross-linkers or interlayer spacers can help prevent restacking, while tailored electrolytes made of ionic liquids or organic solvents can effectively reduce oxidative degradation.

Scalability Challenges: MXenes perform well in the lab, but increasing production in various industrial settings is challenging. Traditional synthesis methods often involve dangerous substances like hydrofluoric acid, raising serious safety and environmental concerns. Maintaining consistency between batches is difficult, partly due to the challenges in controlling surface chemistry and layer thickness. Making electrodes with high mass loading while keeping flexibility and conductivity is complex. Greener, fluoride-free synthesis methods and better process control are needed, along with integration with conductive materials such as graphene or carbon nanotubes to improve mechanical stability.

Cost Challenges: Expensive transition metal precursors and energy-heavy synthesis processes raise the overall costs of producing MXenes. Scaling up often adds expenses while keeping the material quality intact. Incorporating MXenes into existing energy storage systems requires large investments in research and development and significant upgrades to infrastructure. Researchers are now exploring bulk electrochemical exfoliation techniques and hybrid systems that use lower-cost alternatives. These methods balance performance and cost for commercial use.

Strategies to Address Surface Oxidation and Degradation

The stability of MXenes improves significantly through surface passivation or specific chemical modification. Conductive polymers like polyaniline and polypyrrole can coat MXene nanosheets, effectively preventing oxidation while maintaining flexibility and conductivity. Thin oxide coatings such as TiO_2 act as barriers against moisture and air, while functionalizing with organic molecules offers additional resistance. Storing MXenes in inert environments and using fluoride-free synthesis methods can effectively reduce oxidation in some cases. MXene-carbon composites and MXene-MOF hybrids provide extra protection and facilitate swift ion migration through their complex structures. Optimizing electrolytes with the right ionic content or pH can further minimize degradation and extend life during cycling.

Enhancing Stability and Durability of MXene-Based Electrodes

Oxidation, structural degradation, and restacking significantly impact the long-term performance of MXene-based electrodes under various operating conditions. Surface coatings with conductive materials and covalent modifications of surface groups exist, alongside structural engineering methods like 3D porosity or interlayer spacers. The selection of electrolytes is crucial; neutral or redox-active electrolytes may slow down degradation and improve overall charge retention. Strategies such as binder-free electrode fabrication and the integration of MXenes with MOFs and covalent crosslinking strengthen structural integrity and reduce breakdown during demanding cycling processes.

Impact on Performance and Future Prospects

MXenes hold great promise for future energy storage applications. Research is currently focused on detailed structural changes and surface-level engineering to unlock their full potential. Redox-active materials like metal oxides and conductive polymers enhance charge storage when they functionalize MXenes. Combining MXenes with other 2D materials like graphene or MoS_2 prevents restacking and boosts charge transport. Nanoporous MXene structures formed through etching or self-assembly provide better ion accessibility and quicker diffusion. MXene composites combined with materials like MnO_2 or Fe_2O_3 greatly improve conductivity and capacitance for various applications. Single-atom catalysts derived from MXenes are essential for electrocatalysis in rapidly developing hybrid energy systems like water-splitting supercapacitors.

Innovation in electrolytes is also progressing. Ionic liquid and gel polymer electrolytes increase voltage ranges and flexibility, making them ideal for wearable electronics. Redox-active electrolytes (e.g., Fe $(CN)_6^{3-}$, I_3^-) provide additional charge storage through faradaic reactions. Solid-state electrolytes based on MXenes enhance safety and performance further. Sustainable and scalable production is crucial for commercialization. Greener synthesis methods, such as fluoride-free etching and electrochemical exfoliation, are being developed. Roll-to-roll processing and 3D printing enable high-output production of MXene electrodes. Biopolymer composites with cellulose or chitosan also offer environmentally friendly options.

These advancements pave the way for new applications, including flexible and wearable supercapacitors, self-charging hybrid systems, electric vehicles, and battery-supercapacitor hybrids. Overcoming current limitations through material innovation and scalable techniques positions MXenes as vital materials for the future of high-performance and sustainable energy storage systems. Implementing these strategies improves capacitance retention, cycle life, and charge-discharge efficiency. By tackling degradation and structural challenges, MXenes can become more suitable for scalable production and more reliable for real-world energy storage systems. These improvements strengthen their potential for integration into commercial supercapacitor and battery devices.

Conclusion

MXene-based materials exhibit a unique combination of high electrical conductivity, surface tunability, and layered structure, making them exceptional candidates for supercapacitor electrodes. Through advancements in surface engineering, heteroatom doping, and hybridization with redox-active materials, significant improvements have been achieved in capacitance, cycling stability, and charge transport efficiency. Nonetheless, key obstacles—including susceptibility to oxidation, structural restacking, and challenges in scalable synthesis—continue to limit their practical deployment. Efforts to address these limitations through environmentally friendly fabrication methods, electrolyte optimization, and structural stabilization have shown promise. Looking ahead, the integration of MXenes into flexible, wearable, and high-energy storage systems will depend on continued innovation in material processing and device engineering. With sustained research and development, MXenes are well-positioned to play a pivotal role in the future of advanced, sustainable energy storage technologies.

References

[1] P. Naik, M. R. Elmorsy, R. Su, D. D. Babu, A. El-Shafei, and A. V. Adhikari, "New carbazole based metal-free organic dyes with D-π-A-π-A architecture for DSSCs: Synthesis, theoretical and cell performance studies," *Solar Energy*, vol. 153, pp. 600–610, Sep. 2017. https://doi.org/10.1016/J.SOLENER.2017.05.088

[2] K. Zeb et al., "A survey on waste heat recovery: Electric power generation and potential prospects within Pakistan," *Renewable and Sustainable Energy Reviews*, vol. 75, pp. 1142–1155, 2017. https://doi.org/10.1016/j.rser.2016.11.096

[3] Y. Gogotsi and B. Anasori, "The Rise of MXenes," *ACS Nano*, vol. 13, no. 8, pp. 8491–8494, Aug. 2019. https://doi.org/10.1021/ACSNANO.9B06394

[4] D. D. Babu, P. Naik, and K. S. Keremane, "A simple D-A-π-A con fi gured carbazole based dye as an active photo-sensitizer: A comparative investigation on different parameters of cell," *J Mol Liq*, vol. 310, p. 113189, 2020. https://doi.org/10.1016/j.molliq.2020.113189

[5] M. F. Iqbal et al., "Supercapacitors: An Emerging Energy Storage System," *Advanced Energy and Sustainability Research*, p. 2400412, 2025. https://doi.org/10.1002/AESR.202400412

[6] K. Dissanayake and D. Kularatna-Abeywardana, "A review of supercapacitors: Materials, technology, challenges, and renewable energy applications," *J Energy Storage*, vol. 96, p. 112563, Aug. 2024. https://doi.org/10.1016/J.EST.2024.112563

[7] S. Fleischmann et al., "Pseudocapacitance: From Fundamental Understanding to High Power Energy Storage Materials," *Chem Rev*, vol. 120, no. 14, pp. 6738–6782, Jul. 2020. https://doi.org/10.1021/ACS.CHEMREV.0C00170.

[8]P. Simon and Y. Gogotsi, "Materials for electrochemical capacitors," *Nat Mater*, vol. 7, no. 11, pp. 845–854, Nov. 2008. https://doi.org/10.1038/NMAT2297

[9]X. He and X. Zhang, "A comprehensive review of supercapacitors: Properties, electrodes, electrolytes and thermal management systems based on phase change materials," *J Energy Storage*, vol. 56, p. 106023, Dec. 2022. https://doi.org/10.1016/J.EST.2022.106023

[10]   Y. Z. Zhang *et al.*, "MXene hydrogels: fundamentals and applications," *Chem Soc Rev*, vol. 49, no. 20, pp. 7229–7251, Oct. 2020. https://doi.org/10.1039/D0CS00022A

[11]   Y. Gogotsi and Q. Huang, "MXenes: Two-Dimensional Building Blocks for Future Materials and Devices," *ACS Nano*, vol. 15, no. 4, pp. 5775–5780, Apr. 2021. https://doi.org/10.1021/ACSNANO.1C03161

[12]   Y. Kumar *et al.*, "Exploring MXenes and MAX phases: Advancements in properties, synthesis, and application," *Inorg Chem Commun*, vol. 170, p. 113531, Dec. 2024. https://doi.org/10.1016/J.INOCHE.2024.113531

[13]   M. Alhabeb *et al.*, "Guidelines for Synthesis and Processing of Two-Dimensional Titanium Carbide (Ti3C2Tx MXene)," *Chemistry of Materials*, vol. 29, no. 18, pp. 7633–7644, Sep. 2017. https://doi.org/10.1021/ACS.CHEMMATER.7B02847

[14]   A. Hamzehlouy and M. Soroush, "MXene-based catalysts: A review," *Materials Today Catalysis*, vol. 5, p. 100054, Jun. 2024. https://doi.org/10.1016/J.MTCATA.2024.100054

[15]   Y. Cui, J. Zhu, H. Tong, and R. Zou, "Advanced perspectives on MXene composite nanomaterials: Types synthetic methods, thermal energy utilization and 3D-printed techniques," *iScience*, vol. 26, no. 1, p. 105824, Jan. 2023. https://doi.org/10.1016/J.ISCI.2022.105824

[16]   S. Venkateshalu and A. N. Grace, "MXenes—A new class of 2D layered materials: Synthesis, properties, applications as supercapacitor electrode and beyond," *Appl Mater Today*, vol. 18, p. 100509, Mar. 2020. https://doi.org/10.1016/J.APMT.2019.100509

[17]   Q. Yang *et al.*, "MXene/graphene hybrid fibers for high performance flexible supercapacitors," *J Mater Chem A Mater*, vol. 5, no. 42, pp. 22113–22119, Oct. 2017. https://doi.org/10.1039/C7TA07999K

[18]   D. M. Saju, R. Sapna, U. Deka, and K. Hareesh, "MXene material for supercapacitor applications: A comprehensive review on properties, synthesis and machine learning for supercapacitance performance prediction," *J Power Sources*, vol. 647, p. 237302, Aug. 2025. https://doi.org/10.1016/J.JPOWSOUR.2025.237302

[19]   Y. Yu, Q. Fan, Z. Li, and P. Fu, "MXene-based electrode materials for supercapacitors: Synthesis, properties, and optimization strategies," *Materials Today Sustainability*, vol. 24, p. 100551, Dec. 2023. https://doi.org/10.1016/J.MTSUST.2023.100551

[20]   A. M. A, M. Tomy, M. U, and X. T. S, "Supercapacitor featuring Ti3C2Tx MXene electrode: Nanoarchitectonics and electrochemical performances in aqueous and non-aqueous electrolytes," *Mater Res Bull*, vol. 185, p. 113315, May 2025. https://doi.org/10.1016/J.MATERRESBULL.2025.113315

[21]   K. Prenger *et al.*, "Metal Cation Pre-Intercalated Ti3C2TxMXene as Ultra-High Areal Capacitance Electrodes for Aqueous Supercapacitors," *ACS Appl Energy Mater*, vol. 5, no. 8, pp. 9373–9382, Aug. 2022. https://doi.org/10.1021/ACSAEM.2C00653

[22]   M. Hu *et al.*, "Interlayer engineering of Ti3C2Tx MXenes towards high capacitance supercapacitors," *Nanoscale*, vol. 12, no. 2, pp. 763–771, Jan. 2020. https://doi.org/10.1039/C9NR08960H

[23]   T. A. Le *et al.*, "Synergistic Effects of Nitrogen Doping on MXene for Enhancement of Hydrogen Evolution Reaction," *ACS Sustain Chem Eng*, vol. 7, no. 19, pp. 16879–16888, Oct. 2019. https://doi.org/10.1021/ACSSUSCHEMENG.9B04470/SUPPL_FILE/SC9B04470_SI_001.P DF

[24]   Y. Guan *et al.*, "Insight into the mechanism of nitrogen doping in MXenes with controllable surface chemistry," *Mater Today Energy*, vol. 44, p. 101642, Aug. 2024. https://doi.org/10.1016/J.MTENER.2024.101642

[25]   L. Jia *et al.*, "Tuning MXene electrical conductivity towards multifunctionality," *Chemical Engineering Journal*, vol. 475, p. 146361, Nov. 2023. https://doi.org/10.1016/J.CEJ.2023.146361

[26]   A. Shayesteh Zeraati, S. A. Mirkhani, P. Sun, M. Naguib, P. V. Braun, and U. Sundararaj, "Improved synthesis of Ti3C2Tx MXenes resulting in exceptional electrical conductivity, high synthesis yield, and enhanced capacitance," *Nanoscale*, vol. 13, no. 6, pp. 3572–3580, Feb. 2021. https://doi.org/10.1039/D0NR06671K

[27]   R. Akhter and S. S. Maktedar, "MXenes: A comprehensive review of synthesis, properties, and progress in supercapacitor applications," *Journal of Materiomics*, vol. 9, no. 6, pp. 1196–1241, Nov. 2023. https://doi.org/10.1016/J.JMAT.2023.08.011

[28]   C. Zhan, M. Naguib, M. Lukatskaya, P. R. C. Kent, Y. Gogotsi, and D. E. Jiang, "Understanding the MXene Pseudocapacitance," *Journal of Physical Chemistry Letters*, vol. 9, no. 6, pp. 1223–1228, Mar. 2018. https://doi.org/10.1021/ACS.JPCLETT.8B00200/SUPPL_FILE/JZ8B00200_SI_001.PDF

[29]   Z. Li *et al.*, "Intercalation-deintercalation design in MXenes for high-performance supercapacitors," *Nano Res*, vol. 15, no. 4, pp. 3213–3221, Apr. 2022. https://doi.org/10.1007/S12274-021-3939-1

[30]   R. Ibragimova, P. Erhart, P. Rinke, and H. P. Komsa, "Surface Functionalization of 2D MXenes: Trends in Distribution, Composition, and Electronic Properties," *Journal of Physical Chemistry Letters*, vol. 12, no. 9, pp. 2377–2384, Mar. 2021. https://doi.org/10.1021/ACS.JPCLETT.0C03710

[31]   T. Bashir, S. A. Ismail, J. Wang, W. Zhu, J. Zhao, and L. Gao, "MXene terminating groups O, –F or –OH, –F or O, –OH, –F, or O, –OH, –Cl?," *Journal of Energy Chemistry*, vol. 76, pp. 90–104, Jan. 2023. https://doi.org/10.1016/J.JECHEM.2022.08.032

[32]   N. K. Pavithra Siddu, S. M. Jeong, and C. S. Rout, "MXene–carbon based hybrid materials for supercapacitor applications," *Energy Advances*, vol. 3, no. 2, pp. 341–365, Feb. 2024. https://doi.org/10.1039/D3YA00502J

[33]   S. Jayakumar, P. C. Santhosh, M. M. Mohideen, and A. V. Radhamani, "A comprehensive review of metal oxides (RuO2, Co3O4, MnO2 and NiO) for supercapacitor applications and global market trends," *J Alloys Compd*, vol. 976, p. 173170, Mar. 2024. https://doi.org/10.1016/J.JALLCOM.2023.173170

[34]   H. T. Tazwar, M. F. Antora, I. Nowroj, and A. Bin Rashid, "Conductive polymer composites in soft robotics, flexible sensors and energy storage: Fabrication, applications and challenges," *Biosens Bioelectron X*, vol. 24, p. 100597, Aug. 2025. https://doi.org/10.1016/J.BIOSX.2025.100597

[35]   O. Salim, K. A. Mahmoud, K. K. Pant, and R. K. Joshi, "Introduction to MXenes: synthesis and characteristics," *Mater Today Chem*, vol. 14, p. 100191, Dec. 2019. https://doi.org/10.1016/J.MTCHEM.2019.08.010

[36]   M. P. Bilibana, "Electrochemical properties of MXenes and applications," *Advanced Sensor and Energy Materials*, vol. 2, no. 4, p. 100080, Dec. 2023. https://doi.org/10.1016/J.ASEMS.2023.100080

[37]   I. Hussain, A. Hanan, F. Bibi, O. J. Kewate, M. S. Javed, and K. Zhang, "Non-Ti (M2X and M3X2) MXenes for Energy Storage/Conversion," *Adv Energy Mater*, vol. 14, no. 34, p. 2401650, Sep. 2024. https://doi.org/10.1002/AENM.202401650

[38]   R. Ma *et al.*, "Designed Redox-Electrolyte Strategy Boosted with Electrode Engineering for High-Performance Ti3C2Tx MXene-Based Supercapacitors," *Adv Energy Mater*, vol. 13, no. 34, p. 2301219, Sep. 2023. https://doi.org/10.1002/AENM.202301219

[39]   T. Prasankumar *et al.*, "Advancements and approaches in developing MXene-based hybrid composites for improved supercapacitor electrodes," *Materials Today Sustainability*, vol. 28, p. 100963, Dec. 2024. https://doi.org/10.1016/J.MTSUST.2024.100963

[40]   S. Bansal, P. Chaudhary, B. B. Sharma, S. Saini, and A. Joshi, "Review of MXenes and their composites for energy storage applications," *J Energy Storage*, vol. 87, p. 111420, May 2024. https://doi.org/10.1016/J.EST.2024.111420

[41]   S. Panda, K. Deshmukh, S. K. Khadheer Pasha, J. Theerthagiri, S. Manickam, and M. Y. Choi, "MXene based emerging materials for supercapacitor applications: Recent advances, challenges, and future perspectives," *Coord Chem Rev*, vol. 462, p. 214518, Jul. 2022. https://doi.org/10.1016/J.CCR.2022.214518

[42]   M. Hu, M. Zhang, T. Hu, B. Fan, X. Wang, and Z. Li, "Emerging 2D MXenes for supercapacitors: status, challenges and prospects," *Chem Soc Rev*, vol. 49, no. 18, pp. 6666–6693, Sep. 2020. https://doi.org/10.1039/D0CS00175A

Electrocatalysts and Advanced Materials for Sustainable Energy Storage   Materials Research Forum LLC
Materials Research Foundations 182 (2025) 39-56   https://doi.org/10.21741/9781644903797-4

Chapter 4

Recent Developments in the Energy Storage Applications using Polymer Composites

VIPIN Cyriac[1,a], PRADEEP Nayak[1,b], SUDHAKAR Y N[2,c*]

[1]Department of Physics, Manipal Institute of Technology, Manipal Academy of Higher Education, Manipal 576104, Karnataka, India

[2]Department of Chemistry, Manipal Institute of Technology, Manipal Academy of Higher Education, Manipal 576104, Karnataka, India

[a]vipincyriac1729@gmail.com, [b]pradeepnayak1993@gmail.com, [c]sudhakar.yn@manipal.edu

Abstract

Polymer composites are highly efficient and environmentally friendly. These materials not only offer a substantial weight reduction but also exhibit resistance to fatigue and corrosion. They provide excellent "strength to weight" and "stiffness to weight" ratios. Due to their flexibility, cost-effectiveness, and lightness, polymer composites are promising candidates for energy storage applications. However, despite these advantages, they still encounter several challenges. This chapter explores the complexity of composite structures and the relationship between structural parameters. It also addresses the necessity of fillers in polymers for energy storage. The chapter examines the enhancement of conductivity and the types of fillers used in polymer composites, such as carbon-based and metal oxide fillers. Additionally, it covers the applications and recent advancements of polymer composites in supercapacitors. Finally, the chapter provides insights into the challenges and future potential of polymer composites as electrolytes.

Keywords

Polymer Composites, Energy Storage, Supercapacitors, Fillers, Nanocomposites

Contents

Introduction

The demand for advanced energy storage systems has increased in the 21st century due to the worldwide transition towards sustainable energy. Given that renewable energy sources like wind and solar are irregular, there is a need for effective storage to balance supply and demand, as illustrated in Figure 1 [1]. Considering their depletion and effects on the climate, traditional fossil fuels are being phased out. Research on high-performance batteries and supercapacitors for uses such as grid stabilization to electric vehicles is inspired by this context. Supercapacitors, also known as electric double-layer capacitors or EDLCs, have attracted particular interest due to their excellent cycle life, rapid charge and discharge rates, and incredibly high power density. Supercapacitors can deliver power bursts and withstand hundreds of thousands of cycles, but they have a lower energy density than batteries, which have a high energy density of about 150 to 200 Wh/kg but a lower power density [2]. Energy density, power density, cycle life, and rate capability are important performance metrics for supercapacitors. The "holy grail" of energy storage is striking a balance between high energy and power while maintaining a long lifespan.

The performance of supercapacitors is greatly influenced by electrolytes. The internal resistance (equivalent series resistance (ESR)) and operating voltage window are affected directly by the electrolyte's ability to conduct ions between electrodes. Higher cell voltage is made possible by the electrolyte's wider electrochemical stability window, which substantially raises energy density [3]. Similarly, low viscosity and high ionic conductivity reduce ESR, allowing for greater power output. For example, a supercapacitor's energy and power delivery can be greatly increased by using an electrolyte with a wide potential window and strong ionic conductivity. Aqueous solutions like KOH and H_2SO_4 conduct electricity well but are limited to about 1 V because of water splitting [4]. Organic electrolytes, such as tetraethylammonium salts in acetonitrile, allow for 2.5-3 V but are more costly and toxic [5]. Ionic liquids can handle a very high voltage of over 4 V but have lower conductivity and are thicker. Solid-state or gel polymer electrolytes are also common in supercapacitors. Each has trade-offs in terms of environmental impact, safety, ionic conductivity, and voltage range. Despite their historically low conductivity, polymer electrolytes (solid or gel) are becoming more and more appealing for flexible, leak-proof supercapacitors, particularly in wearable applications.

Polymer nanocomposites have become a promising electrolyte material for navigating some of the drawbacks of pure polymers. A polymer composite is defined as a material that integrates a polymer matrix with fillers, specifically nanoscale additives, to enhance its properties. Polymer composites can serve as solid or gel electrolytes in energy storage that are both mechanically strong (no leakage, flexible) and have good ionic transport [6]. The electrolyte's mechanical strength, thermal stability, dielectric constant, and ionic conductivity can all be tuned by adding nanofillers.

Electrocatalysts and Advanced Materials for Sustainable Energy Storage   Materials Research Forum LLC
Materials Research Foundations 182 (2025) 39-56   https://doi.org/10.21741/9781644903797-4

Figure 1. Illustration showing the need for advanced energy storage systems to support integrating intermittent renewables like wind and solar power in the 21st century's transition to sustainable energy.

Two main classes of polymer composites are being studied, which are those based on metal oxides and carbon. Carbon nanostructures, such as graphene, carbon nanotubes, carbon black, and carbon quantum dots, are mixed with the polymer to create carbon-based composites. Because of their high surface area and electrical conductivity, these fillers can enhance the electrolyte's interaction with electrodes and occasionally even produce double-layer capacitance on their own. In metal-oxide-based composites, ceramic nanofillers like TiO_2, ZnO, MnO_2, Fe_3O_4, and NiO are dispersed throughout the polymer. These metal oxides enhance ion mobility by inhibiting polymer crystallization and possessing high dielectric constants [7]. Additionally, some, such as MnO_2 and NiO, are redox-active, potentially leading to pseudocapacitive charge storage.

Each method has benefits and drawbacks. Materials like graphene or carbon nanotubes can create interconnected networks that lower resistance and improve the contact between the electrode and electrolyte. This results in carbon-based polymer electrolytes exhibiting enhanced electronic conductivity and flexibility. Additionally, these materials can also bolster mechanical properties. For instance, the incorporation of a small amount of graphene into a biopolymer hydrogel yielded a specific capacitance of approximately 327 F/g, demonstrating excellent flexibility and enhancement in ionic conductivity to 1.39×10^{-2} S/cm [8]. While aggregation may lead to inconsistent performance, achieving a uniform dispersion of hydrophobic carbon nanomaterials within a polymer matrix presents a significant challenge. Conversely, by enhancing the amorphous fraction of the polymer and providing Lewis acid/base sites that facilitate ion transport, metal oxide-based polymer electrolytes can substantially improve ionic conductivity. For instance, modifications in polymer crystallinity through the incorporation of nano-silica or TiO_2 fillers can result in conductivity enhancements by an order of magnitude [9]. These composites often exhibit superior dielectric strength and thermal stability [10]. However, a challenge arises from the insulating nature of many oxides, as excessive loading may compromise ionic channels or render the electrolyte brittle [11]. Additionally, ensuring electrical contact with the electrode is essential when incorporating pseudocapacitive oxides (such as MnO_2 and Fe_3O_4) into an electrolyte. This task is particularly challenging due to the distribution of the filler within an insulating matrix. In practice, device stability is a critical concern for practical implementation, and numerous studies lack sufficient cycling data on their novel electrolytes. Consequently, both types of composites must ensure compatibility with electrode materials and long-term stability.

In conclusion, polymer nanocomposite electrolytes present an innovative approach to integrating the benefits of nanomaterials, such as high conductivity, permittivity, and additional capacitance, with those of solid polymers, including safety and flexibility. This chapter primarily addresses the laboratory-scale advancements in these composites for supercapacitor electrolytes over the past five years. We specifically analyze the synthesis processes, structure-property relationships, and supercapacitor cell performance of two types: Type 1, carbon-based polymer composites, and Type 2, metal oxide-based polymer composites. Recent case studies involving specific polymers, including bio-derived polymers, are highlighted, with a discussion of their respective advantages and disadvantages. The primary objectives are to elucidate how these polymer nanocomposite electrolytes can enhance energy storage performance and to outline potential future developments for transitioning them from laboratory research to market applications.

Carbon-Based Polymer Composites

In carbon-based polymer composites, carbon nanomaterials are integrated into a polymer electrolyte matrix. Graphene and its oxide (GO), carbon nanotubes (CNTs), activated carbon black, and carbon fibers are examples of common carbon fillers. These fillers can improve the electrolyte's electrochemical interface and potentially add more double-layer capacitance because of their high surface area and exceptional electrical conductivity. This section explores the synthesis of these composites and examines how their structure influences performance.

Synthesis Methods of Carbon-Based Polymer Composites

Carbon-polymer nanocomposite electrolytes are made using different methods. Each method has its own pros and cons. In situ polymerization, solution casting (solution blending), electrospinning, and melt mixing are the main techniques:

(i) In situ polymerization: One popular method for creating carbon-based polymer nanocomposites is in-situ polymerization, which has benefits for improving material properties and dispersing nanofillers. Directly polymerizing monomers on the surface of nanofillers such as carbon nanotubes or graphene oxidimproves interfacial bonding and dispersion [12]. Polyimide, epoxy, and polyurethane are just a few of the polymer systems to which the process can be [13]. It has been demonstrated that highly loaded carbon nanotube-polymer composites with customized polymer distribution and enhanced mechanical properties can be created via in-situ interfacial polymerization [14]. Even at low nanofiller loadings, the resultant nanocomposites show improved thermo-mechanical properties [15]. This technique produces well-dispersed nanotubes with good matrix interaction by grafting polymer chains onto carbon nanotubes without the need for pretreatment [16]. Usually requires several steps or polymerization control; it is difficult to scale because of specific conditions and possible cure viscosity problems.

(ii) Solution casting: To fabricate a composite material, the polymer (or monomer) and carbon nanoparticles are dissolved in a suitable solvent, thoroughly mixed—often employing ultrasonication—and subsequently cast into a film, followed by solvent evaporation. This technique is referred to as solution casting. The primary advantage of this method is its ease of use and adaptability; functionalization of the filler can enhance dispersion within the solution. However, a significant limitation is the necessity for solvent compatibility, as certain fillers or polymers may not disperse effectively and could re-aggregate upon solvent removal. Additionally, the performance of the electrolyte may be adversely affected by residual solvent.

(iii) Electrospinning: is the process of creating nano- to microscale fibers that are gathered into a nonwoven mat by ejecting a polymer solution with distributed carbon nanofillers through a fine needle while applying a high voltage. This can produce fibrous polymer electrolytes with a large surface area and high porosity. To a certain degree, the carbon fillers (such as graphene and carbon nanotubes) can align along the fiber axis, possibly forming conductive pathways [17]. Benefit: Produces thin, porous membranes with superior electrolyte absorption and flexibility. For flexible supercapacitors, it has been utilized to create integrated electrode/electrolyte structures [18]. It can be difficult to control the even distribution of filler in fibers; this calls for volatile solvents and ideal spinning conditions. Without specialized equipment, it is difficult to scale large-area films.

(iv) Melt mixing: One popular method for creating carbon-based polymer nanocomposites, especially those that contain carbon nanotubes (CNTs), is melt mixing. This approach offers benefits in dispersion and scalability by directly integrating nanotubes into polymer matrices via melt processing [19],[20]. It produces more consistent mechanical properties and stable morphologies [21]. Additionally, melt mixing improves the dispersion of carbon nanotubes and the interactions between nanotubes and polymers [19]. The dispersion of carbon nanoparticles in the polymer matrix may not be as good as it might be. Carbon nanofillers can be unreactive with polymer surfaces and tend to aggregate [22].

Structure-Property Relationship of Carbon-Based Polymer Composites

The internal structure and morphology of carbon-based polymer composite electrolytes are closely related to their performance. The kind, size, and dispersion of the carbon filler, the properties of the polymer matrix, and the interactions between the polymer and carbon are important variables.

(i) Conductivity and Morphology: An evenly distributed carbon network can greatly improve the electrolyte's ionic and even electronic pathways. By improving the contact at the electrode-electrolyte interface, a minimal quantity of conductive carbon can reduce internal resistance, despite the fact that the optimal electrolyte is ionically conductive and electronically insulating. For example, a graphene-embedded cotton-based hydrogel electrolyte produced a percolation network that promoted ion transport; the graphene's presence gave charge carriers "smooth pathways," increasing ionic conductivity to 1.39×10^{-2} S/cm (a remarkably high value for a solid electrolyte). The same composite demonstrated how morphology (3D porous network with interconnected graphene) improves performance by achieving a specific capacitance of about 327 F/g and maintaining stability over thousands of cycles [8]. Conversely, inadequate performance arises from the diminished effective conducting surface and ion pathways, which occur when carbon fillers aggregate into large clusters.

(ii) Type, Size, and Dispersion of Filler: The composite material exhibits varied responses to different carbon allotropes. Due to their substantial surface area and aspect ratio, graphene sheets, including graphene oxide, can establish a conductive framework within polymers with minimal weight percentage. The oxygen functional groups present in *graphene oxide (GO)* enhance mechanical strength and compatibility with polar polymers. For example, in an EDLC cell, the incorporation of GO into a PVA/alginate polymer electrolyte resulted in a 96% retention of capacitance after 5000 cycles, whereas the polymer alone may degrade more rapidly [23]. This phenomenon is likely attributed to GO's ability to maintain ionic pathways and reinforce the matrix.Carbon nanotubes (CNTs) can bridge across polymer domains. In addition to improving mechanical modulus (the PVA/CNT composite's Young's modulus is approximately 387 MPa

Electrocatalysts and Advanced Materials for Sustainable Energy Storage   Materials Research Forum LLC
Materials Research Foundations 182 (2025) 39-56   https://doi.org/10.21741/9781644903797-4

compared to a much lower value for pure polymer), a small loading of CNT (e.g., 0.5% in a PVA fiber matrix) increased the composite's dielectric constant and AC conductivity (dielectric $\varepsilon \approx 10.3$ and $\sigma_{ac} \sim 2.7 \times 10^{-7}$ S/m at high frequencies). Although AC conductivity in this context refers to dipolar alignment, it implies enhanced charge transport pathways[24]. This suggests that well-dispersed CNTs can both stiffen the polymer and facilitate charge movement. Although carbon black or activated carbon particles can improve electrode contact and increase the free volume for ion movement, they usually require higher loadings because of their smaller aspect ratio, which can make the electrolyte drier or mechanically weaker because there is less polymer to hold gel.

(iii) Polymer Matrix Influence: Ion transport is significantly influenced by the characteristics of the polymer, such as its polarity, crystallinity, cross-linking, etc. More ion hopping sites and quicker segmental motion are offered by a highly amorphous, polar polymer. For example, poly (ethylene oxide) (PEO) is frequently used in battery electrolytes to dissolve lithium salts; in supercapacitors, poly(vinyl alcohol) (PVA) is frequently utilized in gel electrolytes with KOH or H_2SO_4 because of its capacity to form films and its hydrophilicity. Depending on how they interact, carbon fillers can either increase or decrease the mobility of polymer chains. Nanofillers frequently increase amorphous content and, consequently, ionic conductivity by interfering with polymer crystallization. When ZnO was added to a PVA-cellulose acetate composite, it disrupted the ordered structure of the polymer and lowered the glass transition temperature, allowing ions to flow more freely [25]. Similar to this, graphene or GO sheets in composites can inhibit polymer chain packing, plasticizing the electrolyte, and enhancing ionic conductivity at ideal loadings. Functional groups (-OH, -COOH, -SO3H, etc.) can anchor polymer chains at the filler interface through hydrogen bonds or $\pi-\pi$ interactions with carbon surfaces, particularly GO or functionalized CNT. This interfacial layer exhibits altered dynamics; while new pathways may improve ion transport, excessive binding can immobilize polymer segments and decrease conductivity.

(iv) Polymer-carbon interfacial interactions: Long-term stability hinges on an exceptional interface between the polymer and carbon filler. When interfacial adhesion is strong, the filler remains intact and void-free during cycles of swelling and drying. Research indicates that the functional groups of GO form hydrogen bonds with polymer chains, enhancing mechanical strength [26] and preventing phase separation [27]. A well-bonded interface also helps distribute stress; composite electrolytes in flexible supercapacitors must withstand bending and stretching. For example, a graphene-loaded polymer hydrogel, due to graphene's reinforcement of the polymer matrix, retained about 80% capacitance with good flexibility [28]. Conversely, poor interfacial compatibility can lead to microcracks or delamination during cycling, especially if the device is subjected to flexing or thermal fluctuations.

Structure-property interactions in carbon-based polymer electrolytes show that optimizing composite morphology (filler type, loading, dispersion) improves ionic conductivity, mechanical strength, and device performance. These benefits require homogeneous dispersion of carbon nanofillers and strong polymer interfacial interactions. An ideal filler concentration exists; insufficient amounts provide no percolation benefit, while excess can cause agglomeration or reduced ionic fraction. Scientists aim to identify this optimal point to maximize electrolyte properties while maintaining stability and flexibility.

Case studies: Recent case studies have examined carbon-based polymer composite electrolytes utilized in supercapacitors (Table 1). The primary performance metrics assessed include specific capacitance (Cs), energy density (E), power density (P), and cycle stability.

Table 1. Recent case studies on carbon-based polymer composite electrolytes utilized in supercapacitors.

Polymer Matrix / Carbon Filler	Synthesis / Form Factor	Supercapacitor	Performance	Ref.
Sodium alginate + PVA with graphene oxide (GO) nanofiller – gel polymer electrolyte	*Solution cast* into a membrane; tested in symmetric EDLC cell using carbon cloth	C_s improved; Long cycle life-High C_s~ 633.3 mF cm^{-2} at 0.5 mA cm^{-2} and E=84 mWh cm^{-2} at a high P=500.2 mW cm^{-2} (at 1 mA cm^{-2}). Ion conductivity improved vs. GO-free polymer.	[23]
Cassava starch (CS)/PVA with carbon quantum dots (CDs) – quasi-solid polymer electrolyte (free-standing film)	*Solution casting* of PVA and starch with H_2SO_4 as proton source and carbon dots (0.02 wt%); room-temperature film	C_s=374.6 F g^{-1} (EDLC device); showed 93% capacitance retention after 10,000 cycles. CDs filler enhanced ionic conductivity and reduced cycle degradation. Demonstrates sustainable bio-polymer with carbon additive.	[29]
PVDF-HFP + graphene nanosheets + ionic liquid (IL) – gel polymer electrolyte	*Solution casting* of PVDF-HFP with EMIMBF4 IL, and incorporation of few-layer graphene (exfoliated) for conductivity enhancement	Assembled in EDLC with AC electrodes; Achieved ~80% capacitance retention after 2000 cycles (with one IL); with a different IL, ~89% retention after 10,000 cycles. Graphene increased ionic conductance and helped maintain low ESR over long cycling.	[30]
PVDF-HFP + IL with graphene oxide (GO) – flexible ionogel electrolyte for high-voltage operation	*Solution cast* film from PVDF-HFP, IL, plus GO (few wt%) for mechanical strength and ionic pathway improvement	In a purely capacitive EDLC variant GO-loaded electrolyte kept ~64% after 5000 cycles. The GO filler enabled a thin yet robust film and improved the electrolyte's electrochemical stability window.	[31]
Cotton-derived cellulose hydrogel with graphene (3D composite solid electrolyte)	*In-situ gelation* of cellulose fibers (cotton) with graphene dispersion and ionic solution, forming a 3D hydrogel network	Used in a flexible solid-state SC; ionic conductivity = 13.9×10^{-3} S/cm (very high) due to graphene's pathways. Device showed C_s ~327 F/g (at 3 mV/s) and ~385 F/g at 0.1 A/g, with good cycling stability and flexibility. Graphene provided both mechanical support and enhanced ion transport.	[8]

Research on carbon-based polymer electrolytes shows a common trend. It involves a polymer host, usually a blend, a liquid or ionic liquid for conductivity, and carbon nanofillers to improve conductivity and strength. These combinations have created solid-state supercapacitors that perform almost as well as liquid-electrolyte supercapacitors. The next section will discuss another type of polymer composites, those with metal oxide fillers, which offer different benefits for electrolyte design.

Metal Oxide-Based Polymer Composites

The polymer electrolyte matrix of metal oxide-based polymer composites contains inorganic oxide nanofillers. Many fillers have been studied, including TiO_2, ZnO, MnO_2, Fe_3O_4, NiO, Al_2O_3, SiO_2, and others. Due to their high dielectric constants, these oxides can improve polymer electrolyte ionic conductivity, enhance its mechanical and thermal stability, and occasionally add more electrochemical functionality (such as redox activity in MnO_2 or NiO). After discussing the synthesis of such composites and how their structure influences their properties, this section provides some recent examples.

Synthesis Methods of Metal Oxide-Based Polymer Composites

There are several ways to incorporate metal oxide nanoparticles into a polymer electrolyte, some of which are similar to those for carbon fillers and some of which are unique to oxides.

(i) Sol-gel process: The sol-gel process creates the metal oxide *in situ* within the polymer matrix. A metal alkoxide or salt precursor is combined with the polymer (typically in solution) and allowed to hydrolyse and condense (the sol-gel reaction) to create oxide nanoparticles[32]. For instance, SiO_2 or TiO_2 nanoparticles can be uniformly produced via a sol-gel process in a polymer such as PVDF or PET[33]. Benefit: It is possible to obtain ultra-fine, widely distributed oxide particles (a few nm) that are closely bonded to the polymer matrix. Cons: Chemistry can be complicated; unreacted precursors or leftover solvents may exist. Additionally, the stability of the polymer must be compatible with the reaction conditions (pH, water content).

(ii) Direct blending (solution or melt): Metal oxide nanoparticles (pre-synthesized) can be melted or dispersed in a polymer solution, like solution casting for carbon composites[34]. Agglomerates are frequently broken up by high shear mixing or ultrasonication[35]. Benefit: simplicity of this method- any commercial oxide nanopowder (TiO_2, Al_2O_3, etc.) can be used here. Drawback: Particle aggregation and potential incompatibility make it challenging to achieve homogeneous dispersion (sometimes surface modification of particles is required). Additionally, low viscosity may cause heavy particles to sediment.

(iii) In situ polymerization with oxides: After combining the oxide filler with monomers or oligomers, the polymerization (cross-linking) process is started, trapping the oxide within the developing polymer network. For example, TiO_2 nanoparticles combined with PEGDA (polyethylene glycol diacrylate) monomer can be UV-cured into a solid network with evenly distributed oxide[36]. Benefits include better polymer-filler bonding and good particle distribution, where polymer forms around particles. Cons: The oxide's surface or any absorbed moisture may prevent polymerization; additionally, as the polymer forms, the relative dielectric environment varies, which may impact particle dispersion toward the end of the cure.

(iv) Hydrothermal or solvothermal synthesis and blending: In certain instances, the oxide particles are created hydrothermally either in the presence of the polymer or in a manner that produces a suspension compatible with the polymer. For instance, ZnO or MnO_2 nanostructures could be produced in a mixture with a polymer dissolved in it (which can then be cast into a composite film). Alternatively, a high-surface-area oxide (such as MnO_2 nanoflowers or needle-like ZnO) could be hydrothermally synthesized and then combined with a polymer electrolyte. Benefit: Control over the size and shape of oxide particles (crystalline nanostructures are produced

46

by hydrothermal methods). Cons: An extra step in the particle synthesis process; possible surfactant residues on the particle surface need to be eliminated for optimal ionic conduction.

In situ polymerization and solution blending are two of the most popular metal oxide composite electrolytes. For instance, fumed silica was incorporated into a polymer for a structural supercapacitor using a sol-gel/solution hybrid method: a nano silica-filled polymer gel was cast and tested, achieving good consistency[37]. Sometimes, a porous scaffold technique is employed to incorporate high-loading oxide, such as electrospinning an oxide-containing polymer to create a porous membrane and then infusing liquid electrolytes to create a gel. This was done for a ZnO-based composite that combined the benefits of solid and liquid by soaking electrospun PVA/ZnO fibres in an electrolyte[24].

The selection of the metal oxide also affects synthesis: TiO_2 is available as stable colloids, whereas SiO_2 is easier to disperse and frequently comes surface-treated to be hydrophilic or hydrophobic as needed. ZnO can be challenging because its polar surfaces can aggregate in nonpolar media. Hydrophobic polymers can benefit from surface functionalization (e.g., with silanes) to incorporate oxides (e.g., dispersing ZnO into PVDF requires matching surface energies). The dispersion of nanoparticles (~10–100 nm) in the polymer must be addressed by each technique. Uniformity of ionic conductivity throughout the film and electron microscopy, which checks for particle clustering, are frequently used to confirm good dispersion. Crucially, many metal oxides, such as ZnO and TiO_2, have broad bandgaps and are semiconductors that are frequently hydrophilic, meaning that some water can be retained on their surface. The water must be eliminated from the dry polymer electrolyte to prevent undesired conductivity or reactions. As a result, synthesis frequently involves anhydrous conditions or a drying step.

Using transition metal oxides (MnO_2, NiO, and Fe_3O_4) with multiple valence states requires special attention. These are capable of participating in Faradaic reactions in specific circumstances. For instance, if linked, Fe_3O_4 (mixed Fe(II)/Fe(III) oxide) in a polymer may move electrons, essentially acting in part as an electrode. Instead of carrying electrons between electrodes, which would shorten the device, the filler should be placed in a proper electrolyte to promote ionic movement. As a result, these fillers are usually utilized for electronic conduction below the percolation threshold or when incorporated into a hybrid device for pseudocapacitance. The synthesis techniques ensure that these fillers are dispersed uniformly (to prevent self-discharge by having more conductive particles on one side of the cell than the other). By selecting the right oxide fillers and synthesis pathways, recent advancements have created composite electrolytes that are structural (bearing mechanical loads), high-voltage capable, or flame-retardant. We then go over how these structural elements correspond to properties.

Structure-Property Relationships of Metal Oxide-Based Polymer Composites: The ionic conduction, mechanical strength, and electrochemical behaviour of polymer electrolytes can all be significantly impacted by metal oxide fillers. Crucial details in the relationships between structure and property include:

(i) Ionic conductivity enhancement: The dielectric constants of many metal oxides are high (for example, ε ~80–100 for TiO_2 in rutile form). A high-ε filler increases the number of free charge carriers by dissociating ionic salt aggregates when it is distributed in a polymer-salt matrix. Furthermore, oxide nanoparticles frequently reside at the interfaces between polymer segments, disrupting crystalline domains and increasing the polymer chain's flexibility [38]. Ionic conductivity is significantly increased as a result of the combined effect.

Electrocatalysts and Advanced Materials for Sustainable Energy Storage   Materials Research Forum LLC
Materials Research Foundations 182 (2025) 39-56   https://doi.org/10.21741/9781644903797-4

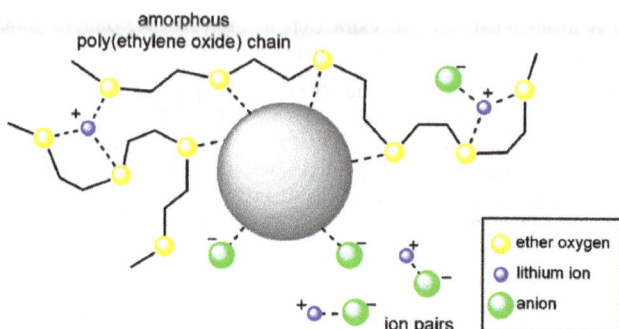

Figure 2. Illustration of Lewis acid-base interactions between the PEO–LiClO₄ polymer
electrolyte matrix and embedded nanoparticles, highlighting their role in modifying the
electrolyte structure and ion transport behavior (Reproduced with permission from Ref. [39],
Copyright 2016, MDPI)

For instance, when 10 weight percent ZnO nanoparticles were added to a PVA–cellulose acetate K⁺-conducting polymer, the room-temperature ionic conductivity rose to 3.70×10^{-3} S/cm, which is almost an order of magnitude higher than the filler-free polymer (10^{-4} S/cm range)[24]. This was explained by the Lewis acid-base interactions between ZnO and the polymer, which promote ion transport and enhanced amorphousness. Likewise, TiO_2 nanoparticles in a PVA-KOH gel electrolyte showed a noticeable increase in ionic conductivity. They enabled the supercapacitor to maintain 100% capacity at 10,000 cycles [40], suggesting that TiO_2 enhanced kinetics and stopped degradative processes, perhaps by scavenging impurities or strengthening the matrix. Building upon these findings, another notable example involves SnO_2 nanoparticles dispersed in polymer electrolytes, as illustrated in Figure 2. The surfaces of SnO_2 particles feature oxygen vacancies that serve as Lewis acid sites, engaging in dual coordination: one with ether oxygen atoms along the polymer backbone, which suppresses crystallinity in poly(ethylene oxide) (PEO) and fosters a higher amorphous phase; and another with the oxygen atoms from the anions of lithium salts such as $LiClO_4$. This latter interaction weakens Li^+–ClO_4^- ion pairing, liberating more free lithium ions for conduction. Both mechanisms synergistically contribute to a marked increase in ionic conductivity in the polymer nanocomposite system.

(ii) Dielectric properties and ion dissociation: Oxide nanofillers increase the composite's dielectric constant (ε'), which aids ionic species dissolution and charge storage. Research shows PVA nanocomposites with ZnO have higher dielectric permittivity than pure PVA at specific frequencies. A higher dielectric constant keeps more ions mobile by reducing coulombic attraction between cation-anion pairs. Surface groups on oxides can coordinate with ions to facilitate transport through a "hopping" pathway. However, an ideal filler content exists, as excess interactions can immobilize ions (10 weight per cent ZnO was optimal, while 15-20 weight per cent reduced conductivity due to particle crowding) [41].

(iii) Polymer-oxide interface and segmental dynamics: Oxide nanoparticles create a tightly bound interfacial layer with the polymer. PVA/CA-ZnO electrolytes' FTIR analysis showed shifts in -OH and C=O vibration bands, indicating coordination bonds between ZnO and polymer chains[25].

The immobilized polymer in this layer can reduce ionic mobility or increase mechanical modulus, with overall effects determined by filler network percolation. When oxides are evenly distributed, polymer regions between particles remain flexible enough to conduct ions, with particle surfaces providing extra pathways. However, when particles clump, they form large, rigid domains that block conduction. Uniform distribution, verified by SEM or TEM imaging, is therefore essential. Two-dimensional oxide fillers like clay can create percolative networks at low loading, altering polymer viscoelastic properties, while spherical metal oxides (TiO_2, ZnO) typically require higher loading for interconnected networks.

(iv) Mechanical and Thermal Stability: Since metal oxide fillers are inorganic, they usually increase the composite's thermal decomposition temperature and modulus. In the polymer matrix, they serve as reinforcements. To improve mechanical properties (such as preventing dendritic penetration in batteries), silica (SiO_2) is frequently added to polymer electrolytes. In supercapacitors, silica-filled polymer electrolytes have allowed the device to function as a load-bearing component. In one instance, a carbon-fiber structural supercapacitor using a composite solid electrolyte with functionalized silica in an epoxy polymer demonstrated 91% capacitance retention after 100 cycles while sustaining mechanical loads[42]. Maintaining electrode spacing and avoiding short circuits under stress or heat can be achieved by reducing the tendency of polymer chains to flow due to the high aspect ratio and hardness of oxide particles.

(v) Surface Area, Porosity, and Crystallinity of Fillers: It is important to consider the oxide filler's actual characteristics rather than just loading. Higher surface area nanoscale oxides offer more interfaces and, consequently, more ion coordination and adsorption locations. Mesoporous TiO_2 particles, for example, may absorb some liquid electrolyte and, in a gel electrolyte, allow ion transport through their pores. The dielectric constant of the filler can be influenced by its crystallinity; crystalline TiO_2 (rutile) has a higher ε than amorphous TiO_2. Additionally, the surface chemistries of different crystal phases vary; for example, anatase TiO_2 may interact with polymers differently than brookite TiO_2. The smaller ~25 nm anatase (P25 TiO_2) produced the best energy storage performance, according to a recent study that compared various TiO_2 nanoparticle sizes/phases (rutile vs. anatase) in a PVA-KOH gel. This was because of its higher surface area and mixed phase, which had synergistic effects[43]. This demonstrates that the performance of the electrolyte is influenced by both the amount and quality of oxide filler.

In conclusion, polymer composites with metal oxides improve ionic conductivity and expand electrochemical stability of polymer electrolytes. They accomplish this by modifying polymer microstructure (increasing amorphous regions and free volume), and utilizing the oxide's dielectric and surface characteristics. The interface between polymers and oxides anchors polymer chains and facilitates ion transport. When properly adjusted, this yields thermally stable, mechanically strong, and highly conductive electrolytes. Although nanoscale fillers only slightly alter flexibility at low loadings, adding inorganic filler can increase weight and reduce electrolyte flexibility.

Case Studies of Recent Advances: We now highlight some recent research on metal oxide–polymer composite electrolytes for supercapacitors, with an emphasis on the composite's composition, manufacturing process, and performance in supercapacitor devices. A comparison of important case studies is given in Table 2.

Electrocatalysts and Advanced Materials for Sustainable Energy Storage   Materials Research Forum LLC
Materials Research Foundations 182 (2025) 39-56   https://doi.org/10.21741/9781644903797-4

Table 2. Recent case studies of metal oxide-based polymer composite electrolytes for supercapacitors.

Polymer Matrix / Oxide Filler	Preparation Method	Supercapacitor Performance / Findings	Ref.
PVA–KOH gel with TiO$_2$ nanoparticles ($\leq$5 wt%)	Simple solution doping of TiO$_2$ (anatase ~20 nm) into PVA–KOH gel; cast into films	Flexible solid-state SC showed 100% capacitance retention after 10,000 cycles – TiO$_2$-doped gel outperformed filler-free gel in stability. High-rate capability improved; attributed to TiO$_2$ raising ionic conductivity and stability. Smaller TiO$_2$ particles (P25, ~21 nm mixed-phase) gave best performance due to higher surface area and favorable phase (anatase/rutile mix)	[40]
PVA/Cellulose Acetate (CA) blend with ZnO nanoparticles (5–15 wt%) and K$_2$CO$_3$ salt	Solution cast of PVA–CA polymer with K$_2$CO$_3$, doped with various ZnO loadings; studied as solid electrolyte	Peak ionic conductivity =3.70×10^{-3} S/cm at 10 wt% ZnO , about tenfold higher than without ZnO. Electrochemical stability window widened to ~3.24 V Filler beyond 15% caused conductivity drop (agglomeration).	[25]
PVDF + ionic liquid with SiO$_2$ nanofiller "ionogel" electrolyte	Solution casting	increase of about 72% in the capacitance compared to a capacitor prepared with the gel electrolyte without SiO2 at ambient temperature.	[44]
PEO-ZnO/rGO/ZnO	Solvothermal process	Improved Ionic Conductivity Enhanced Lithium-Ion Transport. Long term cycling stability of over 900 hours. Specific capacity of 130 mAh/g at 0.5 C after cycling 300 cycles with a poor capacity fading of 0.05% per cycle.	[45]

All of these studies demonstrate that metal oxide–polymer composite electrolytes can be engineered to have mechanical robustness, high conductivity, wide voltage, and even additional charge storage. Based on the intended primary improvement, the type of filler is selected, such as TiO$_2$, Al$_2$O$_3$, and SiO$_2$ for conductivity and stability (inert fillers). Speciality oxides (such as CeO$_2$ for redox shuttle or Fe$_3$O$_4$ for magnetic alignment of electrolyte, etc.) for niche functionalities, redox oxides (MnO$_2$, NiO) for capacitance boosts, and ZnO for conductivity, besides possibly UV-blocking or antimicrobial properties in some cases.

Comparative Analysis and Future Prospects

We can compare the performance characteristics of carbon-based and metal oxide-based polymer nanocomposite electrolytes, consider the practical implications (cost, scalability), discuss new trends that could influence the field's future, and identify research gaps after reviewing both.

Performance Comparison

(i) Ionic conductivity: Metal oxide-filled composites excel at increasing polymer ionic conductivity. Through disrupting polymer crystallinity and creating high-dielectric interfaces, oxides like SiO$_2$, TiO$_2$, and Al$_2$O$_3$ can increase ionic conductivity by an order of magnitude or more[7], [25]. Carbon fillers can also indirectly increase ionic conductivity. For example, graphene oxide (GO) adds polar oxygen groups and inhibits chain crystallization, increasing conductivity. However, purely conductive carbons, such as graphene or carbon nanotubes (CNT), may not greatly increase ion dissociation due to low dielectric constant and may decrease ion movement

volume if overfilled. Carbon additives improve electrode-electrolyte interface wetting and may shorten ion diffusion paths in porous electrodes. In applications, oxide composites typically show greater bulk ionic conductivities than carbon composites at ideal filler loadings.

(ii) Electrochemical Stability (Voltage Window): Metal oxides can scavenge impurities and are electrochemically inert over large voltage ranges, thus increasing the stable voltage window of polymer electrolytes. For instance, ZnO and TiO_2 composites permitted stable operation up to ~3–4 V, while pure aqueous PVA/KOH would be limited to ~1.8 V. Though they don't raise voltage above what the polymer or liquid component permits, carbon materials have wide electrochemical stability, particularly graphitic carbons, stable up to about 4–4.5 V vs. Ag/AgCl in aqueous. Carbon fillers must not add functional groups that might oxidize at high voltage. Therefore, oxide fillers or ionic liquids in polymers have a greater influence on reaching high cell voltage, a path to higher energy density.

(iii) Specific capacitance and Energy density: In EDLCs, electrolytes affect energy density through voltage and ion availability rather than directly influencing capacitance. Double-layer capacitance at the electrode is not altered by carbon or oxide fillers unless they actively participate. Carbon-based composites can increase effective capacitance by enhancing ionic access to pores and potentially adding EDLC from filler contacting electrodes. Inert metal oxide composites won't add EDLC, but can increase capacitance with pseudocapacitive oxides (like MnO_2 or NiO additives, though this enters hybrid device territory). Rather than adding intrinsic capacitance, the main energy benefit of composites for pure EDLC applications is enabling higher voltage and thinner electrolytes (reducing resistance).

(iv) Power Density: The goal of both composites is to increase power by lowering internal resistance. Because conductive carbon reduces interfacial impedance, carbon-based electrolytes perform well in high-power pulses. By increasing ionic conductivity, metal oxide composites also decrease resistance. Carbon fillers may help when the limitation is electronic contact at interfaces, while oxide fillers aid when the limitation is ion transport through electrolyte. Nevertheless, the overall effect on power can be comparable (both permit higher current flow). Both strategies have produced solid-state supercapacitors with high-power densities (~kW/kg levels). While oxide fillers can be added in greater amounts without electronic percolation (they remain insulating), high loadings of carbon may percolate and make the electrolyte electrically conductive (causing leakage current).

(v) Mechanical Flexibility and Strength: By functioning as reinforcing fibers, carbon nanofillers—particularly graphene, or CNT—can significantly increase the toughness and flexibility of polymer electrolytes. Numerous carbon-based composites are said to be flexible and stretchable appropriate for SCs that are wearable. When added in excess, metal oxides can cause the polymer to become more brittle and generally increase stiffness. They can, however, increase strength while maintaining flexibility if used sparingly (in tiny, evenly spaced doses). Therefore, applications requiring extreme flexibility (such as bendable/stretchable supercapacitors) may favor carbon fillers, while those requiring structural rigidity (such as high-temperature stability or structural composites) may favour oxide fillers.

(vi) Thermal Stability: In this case, metal oxide composites are much better than carbon-based composites. The decomposition temperature and frequently the flash point of polymer electrolytes are raised by inorganic fillers. Electrolytes are frequently treated with SiO_2, TiO_2, and other materials to render them non-flammable by absorbing heat and/or releasing endothermic gases.

Since carbon can burn on its own unless the polymer is self-extinguishing, carbon fillers are not very good for flammability. Oxide-filled electrolytes are, therefore more promising for use in extreme environments and for safety.

(vii) Cost and Scalability: Polymer electrolyte composites made with low-cost components are financially attractive. In bulk, metal oxides (like TiO_2, ZnO, and SiO_2) are very inexpensive. There are several carbon fillers. While carbon black is inexpensive, materials like graphene and single-wall carbon nanotubes can be costly. Multi-wall CNTs and graphene oxide are reasonably priced for lab-scale research; however, filler costs may concern large-scale manufacturing. Although graphene prices have decreased, it remains more expensive per kilogram than TiO2. This could work if only a small percentage is needed.

(viii) Processing also affects scalability: Adding fillers doesn't significantly alter the scalability of solution casting as a roll-to-roll coating for polymer films, though industrial mixers and quality control (avoidance of agglomerates) may be necessary to ensure uniform dispersion at high volumes. Melt mixing thermoplastic polymer electrolytes (like PVDF-based ones) with oxide fillers may be preferred because these techniques are directly scalable with current polymer processing equipment (extruders, injection molding); carbon fillers can also handle melt processing, but they can also significantly increase viscosity. Although sensitive chemistry may make in situ sol-gel less scalable, some businesses use sol-gel to create polymer composites on a large scale (e.g. ORMOCERs in coatings).

Conclusion

In conclusion, polymer nanocomposites based on metal oxides and carbon offer strong paths to sophisticated supercapacitor electrolytes. High power performance and integration into flexible devices are facilitated by the exceptional mechanical flexibility and interface provided by carbon-based composites. Oxide of metal-based composites provides improvements in thermal stability, ionic conductivity, and possibly operating voltages. To persuade the industry of their dependability, the present research gaps must be filled, especially in the areas of long-term stability and standardizing performance metrics. Future developments suggest that these composite electrolytes will be used in more multipurpose, printable, and environmentally friendly energy storage devices. The performance gap between solid-state and liquid-electrolyte supercapacitors may be closed by polymer nanocomposite electrolytes with further development, bringing us one step closer to form-flexible, safe, and effective energy storage solutions that are prepared for widespread commercial use.

References

[1] O. I. O., A. S. O., T. A. S., A. I. M., O. O. F., and O. A. O., "Polymer-based nanocomposites for supercapacitor applications: a review on principles, production and products," *RSC Adv*, vol. 15, no. 10, pp. 7509–7534, 2025. https://doi.org/10.1039/D4RA08601E

[2] J. O. Dennis *et al.*, "A Review of Current Trends on Polyvinyl Alcohol (PVA)-Based Solid Polymer Electrolytes," *Molecules*, vol. 28, no. 4, p. 1781, Feb. 2023. https://doi.org/10.3390/molecules28041781

[3] A. Mendhe and H. S. Panda, "A review on electrolytes for supercapacitor device," *Discov Mater*, vol. 3, no. 1, p. 29, Oct. 2023. https://doi.org/10.1007/s43939-023-00065-3

[4]C. Zhao and W. Zheng, "A Review for Aqueous Electrochemical Supercapacitors," *Front Energy Res*, vol. 3, May 2015. https://doi.org/10.3389/fenrg.2015.00023

[5]S. Biswas and A. Chowdhury, "Organic Supercapacitors as the Next Generation Energy Storage Device: Emergence, Opportunity, and Challenges," *ChemPhysChem*, vol. 24, no. 3, Feb. 2023. https://doi.org/10.1002/cphc.202200567

[6]X. Chen and R. Holze, "Polymer Electrolytes for Supercapacitors," *Polymers (Basel)*, vol. 16, no. 22, p. 3164, Nov. 2024. https://doi.org/10.3390/polym16223164

[7]S. Jayanthi *et al.*, "The Transformative Role of Nano-SiO2 in Polymer Electrolytes for Enhanced Energy Storage Solutions," *Processes*, vol. 12, no. 10, p. 2174, Oct. 2024. https://doi.org/10.3390/pr12102174

[8]N. B. Mohammed *et al.*, "Natural Solid-State Hydrogel Electrolytes Based on 3D Pure Cotton/Graphene for Supercapacitor Application," *Micromachines (Basel)*, vol. 14, no. 7, p. 1379, Jul. 2023. https://doi.org/10.3390/mi14071379

[9]S. Jayanthi *et al.*, "The Transformative Role of Nano-SiO2 in Polymer Electrolytes for Enhanced Energy Storage Solutions," *Processes*, vol. 12, no. 10, p. 2174, Oct. 2024. https://doi.org/10.3390/pr12102174

[10]　V. R. Jeedi, K. K. Ganta, I. S. R. Varma, M. Yalla, S. Narender Reddy, and A. S. Chary, "Alumina Nanofiller Functionality on Electrical and Ion Transport Properties of PEO-PVdF/KNO3/SN Nanocomposite Polymer Electrolytes," *Results Chem*, vol. 5, p. 100814, Jan. 2023. https://doi.org/10.1016/j.rechem.2023.100814

[11]　K. Suhailath, M. T. Ramesan, B. Naufal, P. Periyat, V. C. Jasna, and P. Jayakrishnan, "Synthesis, characterisation and flame, thermal and electrical properties of poly (n-butyl methacrylate)/titanium dioxide nanocomposites," *Polymer Bulletin*, vol. 74, no. 3, pp. 671–688, Mar. 2017. https://doi.org/10.1007/s00289-016-1737-9

[12]　Ph. Dubois and M. Alexandre, "Performant Clay/Carbon Nanotube Polymer Nanocomposites," *Adv Eng Mater*, vol. 8, no. 3, pp. 147–154, Mar. 2006. https://doi.org/10.1002/adem.200500256

[13]　H. Mao and X. Wang, "Use of in-situ polymerization in the preparation of graphene / polymer nanocomposites," *New Carbon Materials*, vol. 35, no. 4, pp. 336–343, Jul. 2020. https://doi.org/10.1016/S1872-5805(20)60493-0

[14]　C. A. C. Chazot, C. K. Jons, and A. J. Hart, "In Situ Interfacial Polymerization: A Technique for Rapid Formation of Highly Loaded Carbon Nanotube-Polymer Composites," *Adv Funct Mater*, vol. 30, no. 52, Dec. 2020. https://doi.org/10.1002/adfm.202005499

[15]　Ph. Dubois and M. Alexandre, "Performant Clay/Carbon Nanotube Polymer Nanocomposites," *Adv Eng Mater*, vol. 8, no. 3, pp. 147–154, Mar. 2006. https://doi.org/10.1002/adem.200500256

[16]　G. Viswanathan *et al.*, "Single-Step in Situ Synthesis of Polymer-Grafted Single-Wall Nanotube Composites," *J Am Chem Soc*, vol. 125, no. 31, pp. 9258–9259, Aug. 2003. https://doi.org/10.1021/ja0354418

[17]   L. Y. Yeo and J. R. Friend, "Electrospinning carbon nanotube polymer composite nanofibers," *J Exp Nanosci*, vol. 1, no. 2, pp. 177–209, Jun. 2006. https://doi.org/10.1080/17458080600670015

[18]   Y.-E. Miao, J. Yan, Y. Huang, W. Fan, and T. Liu, "Electrospun polymer nanofiber membrane electrodes and an electrolyte for highly flexible and foldable all-solid-state supercapacitors," *RSC Adv*, vol. 5, no. 33, pp. 26189–26196, 2015. https://doi.org/10.1039/C5RA00138B

[19]   O. Breuer and U. Sundararaj, "Big returns from small fibers: A review of polymer/carbon nanotube composites," *Polym Compos*, vol. 25, no. 6, pp. 630–645, Dec. 2004. https://doi.org/10.1002/pc.20058

[20]   P. Pötschke *et al.*, "Melt Mixing as Method to Disperse Carbon Nanotubes into Thermoplastic Polymers," *Fullerenes, Nanotubes and Carbon Nanostructures*, vol. 13, no. sup1, pp. 211–224, Apr. 2005. https://doi.org/10.1081/FST-200039267

[21]   C. Lee and M. D. Dadmun, "Improving the dispersion and interfaces in polymer-carbon nanotube nanocomposites by sample preparation choice," *J Polym Sci B Polym Phys*, vol. 46, no. 16, pp. 1747–1759, Aug. 2008. https://doi.org/10.1002/polb.21510

[22]   J.-V. Lim, S.-T. Bee, L. Tin Sin, C. T. Ratnam, and Z. A. Abdul Hamid, "A Review on the Synthesis, Properties, and Utilities of Functionalized Carbon Nanoparticles for Polymer Nanocomposites," *Polymers (Basel)*, vol. 13, no. 20, p. 3547, Oct. 2021. https://doi.org/10.3390/polym13203547

[23]   M. Najafloo and L. Naji, "Resilient 3D porous self-healable triple network hydrogels reinforced with graphene oxide for high-performance flexible supercapacitors," *J Alloys Compd*, vol. 1002, p. 175235, Oct. 2024. https://doi.org/10.1016/j.jallcom.2024.175235

[24]   S. A. Shah, H. Ali, M. I. Inayat, E. E. Mahmoud, H. AL Garalleh, and B. Ahmad, "Effect of carbon nanotubes and zinc oxide on electrical and mechanical properties of polyvinyl alcohol matrix composite by electrospinning method," *Sci Rep*, vol. 14, no. 1, p. 28107, Nov. 2024. https://doi.org/10.1038/s41598-024-79477-x

[25]   J. Ojur Dennis *et al.*, "Effect of ZnO Nanofiller on Structural and Electrochemical Performance Improvement of Solid Polymer Electrolytes Based on Polyvinyl Alcohol–Cellulose Acetate–Potassium Carbonate Composites," *Molecules*, vol. 27, no. 17, p. 5528, Aug. 2022. https://doi.org/10.3390/molecules27175528

[26]   C. Poochai *et al.*, "Alpha-MnO2 nanofibers/nitrogen and sulfur-co-doped reduced graphene oxide for 4.5 V quasi-solid state supercapacitors using ionic liquid-based polymer electrolyte," *J Colloid Interface Sci*, vol. 583, pp. 734–745, Feb. 2021. https://doi.org/10.1016/j.jcis.2020.09.045

[27]   X. Yang, F. Zhang, L. Zhang, T. Zhang, Y. Huang, and Y. Chen, "A High-Performance Graphene Oxide-Doped Ion Gel as Gel Polymer Electrolyte for All-Solid-State Supercapacitor Applications," *Adv Funct Mater*, vol. 23, no. 26, pp. 3353–3360, Jul. 2013. https://doi.org/10.1002/adfm.201203556

[28]   N. B. Mohammed *et al.*, "Natural Solid-State Hydrogel Electrolytes Based on 3D Pure Cotton/Graphene for Supercapacitor Application," *Micromachines (Basel)*, vol. 14, no. 7, p. 1379, Jul. 2023. https://doi.org/10.3390/mi14071379

[29]   T. Jorn-am, P. Supchocksoonthorn, W. Pholauyphon, J. Manyam, C. Chanthad, and P. Paoprasert, "Quasi-Solid, Bio-Renewable Supercapacitors Based on Cassava Peel and Cassava Starch and the Use of Carbon Dots as Performance Enhancers," *Energy & Fuels*, vol. 36, no. 14, pp. 7865–7877, Jul. 2022. https://doi.org/10.1021/acs.energyfuels.2c01263

[30]   M.-J. Shi, S.-Z. Kou, B.-S. Shen, J.-W. Lang, Z. Yang, and X.-B. Yan, "Improving the performance of all-solid-state supercapacitors by modifying ionic liquid gel electrolytes with graphene nanosheets prepared by arc-discharge," *Chinese Chemical Letters*, vol. 25, no. 6, pp. 859–864, Jun. 2014. https://doi.org/10.1016/j.cclet.2014.04.010

[31]   X. Yang, F. Zhang, L. Zhang, T. Zhang, Y. Huang, and Y. Chen, "A High-Performance Graphene Oxide-Doped Ion Gel as Gel Polymer Electrolyte for All-Solid-State Supercapacitor Applications," *Adv Funct Mater*, vol. 23, no. 26, pp. 3353–3360, Jul. 2013. https://doi.org/10.1002/adfm.201203556

[32]   A. Rajani, T. Singh Anand, and P. Dave, "REVIEW ON SYNTHESIS OF METAL DOPED METAL OXIDE NANOCOMPOSITES BY SOL-GEL METHOD TO EXAMINE PHOTO-CATALYTIC ACTIVITY TO DISTINGUISH DIFFERENT ORGANIC-INORGANIC CONTAMINANTS," *International Journal of Creative Research Thoughts*, vol. 8, pp. 2320–2882, 2020, Accessed: May 28, 2025. [Online]. Available: www.ijcrt.org

[33]   Q. Ji, X. Wang, Y. Zhang, Q. Kong, and Y. Xia, "Characterization of Poly (ethylene terephthalate)/SiO2 nanocomposites prepared by Sol–Gel method," *Compos Part A Appl Sci Manuf*, vol. 40, no. 6–7, pp. 878–882, Jul. 2009. https://doi.org/10.1016/j.compositesa.2009.04.010

[34]   A. Moghadam, M. Salmani Mobarakeh, M. Safaei, and S. Kariminia, "Synthesis and characterization of novel bio-nanocomposite of polyvinyl alcohol-Arabic gum-magnesium oxide via direct blending method," *Carbohydr Polym*, vol. 260, p. 117802, May 2021. https://doi.org/10.1016/j.carbpol.2021.117802

[35]   C. Dossin Zanrosso, D. Piazza, and M. A. Lansarin, "PVDF/ZnO composite films for photocatalysis: A comparative study of solution mixing and melt blending methods," *Polym Eng Sci*, vol. 60, no. 6, pp. 1146–1157, Jun. 2020. https://doi.org/10.1002/pen.25368

[36]   N. A. Slesarenko *et al.*, "Nanocomposite Polymer Gel Electrolyte Based on TiO2 Nanoparticles for Lithium Batteries," *Membranes (Basel)*, vol. 13, no. 9, p. 776, Sep. 2023. https://doi.org/10.3390/membranes13090776

[37]   D. W. Kim, S. M. Jung, and H. Y. Jung, "A super-thermostable, flexible supercapacitor for ultralight and high performance devices," *J Mater Chem A Mater*, vol. 8, no. 2, pp. 532–542, 2020. https://doi.org/10.1039/C9TA11275H

[38]   B. Scrosati, F. Croce, and L. Persi, "Impedance Spectroscopy Study of PEO-Based Nanocomposite Polymer Electrolytes," *J Electrochem Soc*, vol. 147, no. 5, p. 1718, 2000. https://doi.org/10.1149/1.1393423

[39]   W. Wang and P. Alexandridis, "Composite Polymer Electrolytes: Nanoparticles Affect Structure and Properties," *Polymers (Basel)*, vol. 8, no. 11, p. 387, Nov. 2016. https://doi.org/10.3390/polym8110387

[40]   K. Wongsaprom, P. Insee, N. Boonraksa, and E. Swatsitang, "Enhancement of electrochemical performance in supercapacitors using TiO2 nanoparticles-doped PVA-KOH gel electrolyte," *Journal of Electroanalytical Chemistry*, vol. 978, p. 118886, Feb. 2025. https://doi.org/10.1016/j.jelechem.2024.118886

[41]   Y. M. Volfkovich, A. Y. Rychagov, V. E. Sosenkin, S. A. Baskakov, E. N. Kabachkov, and Y. M. Shulga, "Supercapacitor Properties of rGO-TiO2 Nanocomposite in Two-component Acidic Electrolyte," *Materials*, vol. 15, no. 21, p. 7856, Nov. 2022. https://doi.org/10.3390/ma15217856

[42]   M. Hwang *et al.*, "Composite solid polymer electrolyte with silica filler for structural supercapacitor applications," *Korean Journal of Chemical Engineering*, vol. 38, no. 2, pp. 454–460, Feb. 2021. https://doi.org/10.1007/s11814-020-0695-y

[43]   T.-R. Kuo, L.-Y. Lin, K.-Y. Lin, and S. Yougbaré, "Effects of size and phase of TiO2 in poly (vinyl alcohol)-based gel electrolyte on energy storage ability of flexible capacitive supercapacitors," *J Energy Storage*, vol. 52, p. 104773, Aug. 2022. https://doi.org/10.1016/j.est.2022.104773

[44]   P. F. R. Ortega, J. P. C. Trigueiro, G. G. Silva, and R. L. Lavall, "Improving supercapacitor capacitance by using a novel gel nanocomposite polymer electrolyte based on nanostructured SiO2, PVDF and imidazolium ionic liquid," *Electrochim Acta*, vol. 188, pp. 809–817, Jan. 2016. https://doi.org/10.1016/j.electacta.2015.12.056

[45]   W. Zhang *et al.*, "Sandwich Structured Metal oxide/Reduced Graphene Oxide/Metal Oxide-Based Polymer Electrolyte Enables Continuous Inorganic–Organic Interphase for Fast Lithium-Ion Transportation," *Small*, vol. 19, no. 19, May 2023. https://doi.org/10.1002/smll.202207536

Electrocatalysts and Advanced Materials for Sustainable Energy Storage   Materials Research Forum LLC
Materials Research Foundations 182 (2025) 57-69   https://doi.org/10.21741/9781644903797-5

Chapter 5

Hybrid Electrocatalysts: Next-Generation Composites for Storing Energy in a Sustainable Way

ATHIRA Suresh[1,a], DIVYA Sreetha Murugan[1,b], BIJI Pullithadathil[1,c*]

[1]Department of Chemistry & Nanoscience and Technology, PSG Institute of Advanced Studies, Coimbatore-641004, India

[a]ash@psgias.ac.in, [b]mds@psgias.ac.in, [c]pbm@psgias.ac.in

Abstract

The thrust for effective energy storage and conversion technologies has been fueled by the growing demand for clean and renewable energy sources. There is immense potential for developing these technologies with the help of hybrid electrocatalysts, which are composite materials composed of metals, metal oxides, carbon-based compounds, and polymers. The key electrochemical reactions that are essential for energy storage and fuel cell applications, including the oxygen evolution reaction (OER), hydrogen evolution reaction (HER), Oxygen reduction reaction (ORR) and CO_2 reduction, are improved by these materials. By leveraging the unique characteristics of each component, hybrid electrocatalysts maximize their catalytic efficiency, stability, conductivity and selectivity while minimizing the need for costly or rare elements like platinum. For example, carbon-based materials offer improved electrical conductivity and durability, metals offer high catalytic activity, and metal oxide enhances stability and corrosion resistance. These materials can be combined to generate hybrid electrocatalysts, which can overcome the drawbacks of individual elements and produce more economical and effective fuel cells, supercapacitors, and hydrogen generation systems. Despite the promising advantages of hybrid electrocatalysts, there still exist challenges such as materials scalability cost, and long-term stability. Hybrid electrocatalysts are anticipated to be essential to the creation of efficient and economically feasible sustainable energy storage and conversion technology.

Keywords

Hybrid-Electrocatalyst, Energy Storage, Synergistic Effect, Renewable Energy, Catalyst Stability

Contents

Introduction

The growing need for fossil fuel alternatives and the rising energy demand have spurred researchers all over the world to actively investigate new sustainable energy sources and technologies. Economic growth demands high energy and environmental consumption. In order to address the energy crisis and environmental pollution issues, hydrogen (H_2), with an effective electrocatalyst, is a viable alternative to fossil fuels as a clean and abundant energy source [1]. Electrocatalysts are crucial materials that accelerate key electrochemical reactions such as the hydrogen evolution reaction (HER), oxygen reduction reaction (ORR), and carbon dioxide reduction reaction (CO_2RR), all of which are fundamental to sustainable energy technologies. In HER, electrocatalysts facilitate the production of hydrogen gas from water, making clean hydrogen generation more efficient [2]. In ORR, which occurs in fuel cells and metal-air batteries, catalysts help reduce oxygen molecules into water or hydroxide ions, significantly improving energy conversion efficiency [3]. Wang et al., demonstrated the high-performance universal ORR electrocatalysts in both basic and acidic media by synthesizing the $WCFeWO_4@FeNOMC$ hybrid. This hybrid may be an attractive Non-Precious Metals catalyst for ORR because it has a desirable ORR activity, prior stability and an excellent tolerance capability to CH_3OH when compared to commercial Pt/C catalyst, which demonstrates that the routes to innovative energy materials are expanded by the synergistic action of nanomaterials [4]. Meanwhile, in CO_2RR, electrocatalysts enable the transformation of carbon dioxide into value-added chemicals and fuels like CO, formic acid, or hydrocarbons, offering a promising route for carbon recycling and renewable fuel production. Each reaction requires specific catalysts to overcome kinetic barriers, reduce overpotentials, and achieve high selectivity, making the design and optimization of electrocatalysts a central focus in clean energy research. The best catalysts for the hydrogen evolution reaction (HER) are thought to be Pt-based. However, their industrialization uses are limited by their high cost. A lot of work was carried out to create extremely effective HER electrocatalysts, which were intended to replace Pt-based catalysts. Given its high-power density, high stability, and environmental friendliness, electrochemical energy conversion technology is regarded as one of the most promising options among the many available. Furthermore, some electrochemical energy conversion technologies, such as fuel cells can also deliver high energy densities, making them viable solutions for the creation of sustainable and green energy.

Hybrid electrocatalysts are materials that combine two or more different types of substances to enhance the overall performance for electrochemical reactions, particularly in processes like fuel cells, water splitting, and other energy conversion/storage technologies. The primary goal of hybrid electrocatalyst design is to improve catalytic activity, stability, and efficiency by leveraging the unique properties of each individual component. Platinum (Pt), Palladium (Pd), Rhodium (Rh), and Gold (Au) are commonly used for their high catalytic activity, especially in reactions such as ORR, HOR, and HER. However, their high cost and limited availability drive the search for alternatives or ways to use them more efficiently in hybrid forms. For example, metal oxides like iridium oxide, ruthenium oxide, Carbon-based materials like graphene, carbon nanotubes, for stability or electron transfer facilitation polymer materials are frequently used. The combination of these materials is designed to enhance catalytic activity by optimizing electronic properties, such as charge transfer and activation energy and also to improve stability under harsh conditions (e.g., acidic or alkaline environments). Perivoliotis et al. have reviewed metal-based hybrids supported on graphene, and comparable two-dimensional materials have recently been investigated as possible oxygen reduction reaction electrocatalysts (Figure 1). Reducing the cost by replacing expensive metals with more abundant alternatives can increase conductivity and promote more efficient electron and ion transport.

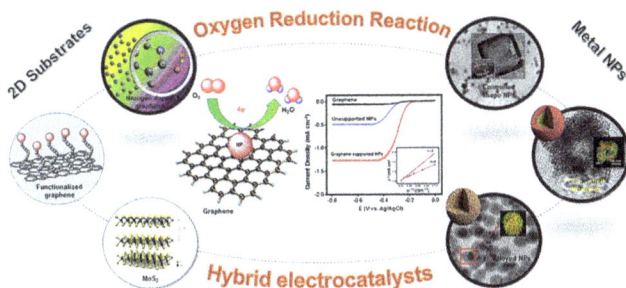

Figure 1. An Overview of metal-based hybrids supported on graphene and related 2D materials used as potential hybrid electrocatalysts (Reproduced with permission from Ref.[6], Copyright 2017, Elsevier).

Yang et al. developed a Co-based MOF (ZIF-67) proposed structure that was inserted by carbon nanotubes (CNTs) and used as a template to create a hybrid of hollow Co nanoparticles (NPs). CNT-interconnected encapsulated N-doped porous carbon nano frames (referred to as Co-NC/CNT). With a cell voltage of 1.625 V to provide a current density of 10 mA cm^{-2}, the resulting continuous 3D network of Co-NC/CNT hybrid, which has highly exposed active sites and good conductivity, demonstrates high bifunctional catalytic performance toward fast water splitting in alkaline solution [5].

Hybrid Electrocatalyst: Composition and Design

For a certain catalytic reaction, the ultimate goal of catalyst design is always to design and produce catalytic materials with high activity, selectivity, and stability. One is that catalysts have developed from homogeneous single-component structures to hybrid multicomponent forms. The primary driving force is that the integration of multiple components creates a synergistic effect, enhancing catalytic performance and enabling the possibility of compensating for each component's deficit. While the other components support the operation of surface catalysis, such as charge production in electrocatalysis and charge transfer, certain components in hybrid structures can offer active sites for surface reactions with high molecular activation ability [7]. The other trend is that improving the performance of electrocatalytic materials has become more and more dependent on the surface and interface design of the materials [8]. The design of hybrid electrocatalysts revolves around several principles, such as the materials being chosen to interact synergistically, meaning that the combination of materials leads to better performance than individual components. For example, Carbon-based supports (e.g., graphene or carbon nanotubes) can provide a high surface area and conductivity, while metallic nanoparticles dispersed on the surface can serve as active catalytic sites. Metal oxides also may serve as promoters, providing high surface areas and redox-active sites for reactions, while metals (like Pt) can improve the efficiency of reaction rates. In recent days, hybrid systems often involve the tuning of the electronic properties of one material (like doping metals with non-metals) to enhance the catalytic activity of the other, and also the shape, size, and distribution of catalytic materials on the hybrid support are critical. For instance, small nanoparticles of metals may be dispersed over a carbon matrix to maximize surface area and catalytic sites while minimizing metal usage.

Hybrid catalysts can be designed to promote interactions between different phases, such as between a metal and an oxide, or between a metal and a conductive carbon structure. This promotes better charge and mass transfer during the reaction, which enhances overall catalytic efficiency. Several materials are commonly employed in hybrid electrocatalyst design [9]. Platinum (Pt), Palladium (Pd), Rhodium (Rh), and Gold (Au) are commonly used for their high catalytic activity, especially in reactions such as ORR, HOR, and hydrogen evolution reaction (HER). However, their high cost and limited availability drive the search for alternatives or ways to use them more efficiently in hybrid forms. Transition metal oxides (e.g., IrO_2, RuO_2, and Co_3O_4) offer high catalytic activity and are often used in OER and ORR, especially in acidic and alkaline electrolytes. When combined with carbon-based supports or other metals, their stability and catalytic performance can be significantly improved. Transition metal sulfides, nitrides, and phosphides are emerging as promising candidates for electrochemical reactions due to their low cost and high catalytic activity [10].

Materials and Strategies for the Design of Hybrid Electrocatalysts

Hybrid electrocatalysts combine different types of materials, often inorganic and organic to enhance the efficiency, selectivity, and stability of electrocatalytic reactions, such as water splitting, CO_2 reduction, and fuel cell reactions. These catalysts often leverage the complementary properties of different components to optimize performance.

Metal Carbon based Hybrid Electrocatalyst: Carbon-based materials such as graphene, carbon nanotubes, or porous carbon materials are often used to support metal catalysts like platinum (Pt), gold (Au), or palladium (Pd). Carbon materials can provide a high surface area, electron conductivity, and stability, while metals serve as the active catalytic sites. Doping of carbon materials with heteroatoms (e.g., nitrogen or boron) further improves electronic conductivity and catalytic properties for reactions like hydrogen evolution. Carbon based hybrid electrocatalyst combines high catalytic activity of metals with conductivity, stability and high surface area of carbon materials, often leading to enhanced performance, durability and cost-efficiency. For improving the ORR Catalytic activity design of transition metal-heteroatom-C electrocatalyst is highly promising. In recent years, Xiao et al. demonstrated that using Fe-N-C as a carbon substrate, a Pt-Fe-N-C hybrid electrocatalyst comprising atomically distributed Pt and Fe single atoms and Pt-Fe alloy nanoparticles was effectively created. The study shows that ultra-low Pt doping can effectively increase the durability of Fe N-C material. which emphasizes the significance of the strong contact between active sites and carbon support as well as the synergy effect between Pt and Fe single atoms, which may provide insight into enhancing the robustness of non-precious metal catalysts through a hybrid structure [11].

Jha et al., synthesized three-dimensional PtRu/Pt nanoparticles that could be decorated over a mixture of one-dimensional f-MWNT and two-dimensional f-G in an equal amount as catalyst support material to reach a maximum power density of roughly 68 mW/cm^2 in DMFC. He demonstrated that by preventing the restacking of exfoliated graphene sheets, PtRu/Pt nanoparticles and f-MWNT may increase the surface area available for catalyst loading. For the methanol oxidation reaction in DMFC, they have shown that PtRu/(f-Gef-MWNT) hybrid nanomaterials exhibit superior electrocatalytic activity in comparison to either PtRu/f-MWNT or PtRu/f-G [12]. Arici et al., synthesized Pt/Graphene nanoplates and Pt/Carbon Black-Graphene nanoplates hybrid electrocatalysts are the first in literature to be reported with the organometallic for PEM fuel cell application. Hybrid structure limited the restacking of GO layers, effectively modified the array of graphene, and sustained more accessible active sites for the fuel cell reactions. He concluded that the incorporation of graphene and its derivatives achieves optimum hybrid support materials and decreases the amount of Pt for better PEM fuel cell performances [13].

Metal Oxide based Hybrid Electrocatalyst: Metal oxides such as IrO_2, RuO_2, or Co_3O_4 are often combined with metals like platinum or palladium to enhance electrochemical stability and activity for reactions such as oxygen evolution (OER) and oxygen reduction (ORR). Metal oxide supports can offer excellent chemical stability in acidic and alkaline conditions, making them ideal for electrolysis and fuel cell applications. Development of highly active, stable, and reasonably priced HER electrocatalysts is a crucial step in achieving a high-efficiency water splitting process. In recent years, metal oxide-based materials have become desirable choices for catalyzing HER because of their improved structural and compositional strengths. Zhu et.al have discussed the metal oxide based materials as an emerging material for HER. In that he has discussed some of the recent advancements in the field of metal oxide based HER, like single transition metal oxides, spinel oxides, perovskite oxides, metal (oxy)hydroxides, specially-structured metal oxides and oxide-containing hybrids. Compared to a single-phase metal oxide catalyst, mixed metal oxide (MMO) electrocatalysts emerge as a potential HER catalyst, providing heterostructured interfaces that can overcome the activation barrier effectively for the hydrogen evolution reaction [14].

Pratama et.al in their article have discussed MMO based electrocatalysts. where she explained that enhancing catalytic activity is largely dependent on the heterojunction on the metal oxide/metal oxide interface; the heterojunction in the heterostructured MMO has demonstrated superior HER activity compared to alloys or oxide composites. Compared to alloys or oxide composites, MMOs' active site was more exposed by the heterojunction [15]. Wei and colleagues created NiO/Co_3O_4 concave surface microcubes from a metal–organic framework (MOF) precursor ($Ni_3[Co(CN)_6]_2$) to enhance the number of heterojunctions to an atomic level by adding oxygen vacancies to the catalyst [16]. Singh et al. were able to increase the HER activity of a TiO_2/ZrO_2 composite and achieve a Tafel slope value of 87 mV dec^{-1} and an overpotential value of 160 mV at 10 mA cm^{-2}. A charge imbalance on the heterostructure could be caused by the ZrO_2-induced grain boundary defects. Consequently, an additional proton attaches itself to the oxygen atom linked to the charge imbalance, generating an excess of -OH groups on the surface and raising the acidity of the surface [17].

Core-shell based Hybrid Electrocatalyst: Core-shell based electrocatalysts are advanced materials. These electrocatalysts consist of a core material surrounded by a shell. The core is typically made of a less expensive or non-noble metal, while the shell is composed of a noble metal like platinum or palladium. This design reduces the use of costly noble metals while maintaining high catalytic activity. These catalysts offer improved activity, stability, and resistance to chemical erosion. They also mitigate issues like nanoparticle aggregation and sintering, which can degrade performance over time. Guo et al. explored a new method for creating an Au/Pt hybrid electrocatalyst with a sponge-like and hollow structure that has been effectively developed. This specific architecture, which is hollow and sponge-like, is ideal for catalytic applications due to its low cost and great efficiency. Initially, this unique hybrid nanomaterial's potential as an anodic and cathodic electrocatalyst was examined. The hybrid nanomaterial outperforms the standard Pt electrode in terms of catalytic activity for methanol oxidation and dioxygen reduction. The findings have implications for fuel cells, sensors, electrocatalysis, and shape-controlled production of hybrid nanomaterials [18].

Magro et al., demonstrated that the final product was a core-shell structure made up of cubic nanostructures of maghemite nanoparticles covered in crystalline Prussian blue by simply incubating Surface Active Maghemite Nanoparticles in water containing ferrocyanide, the material was readily synthesized and used to create an electrode intended for hydrogen peroxide electro-reduction. The suggested electrode expands the range of applications for Prussian blue as an electrocatalyst and may effectively compete with other H_2O_2 detection systems because of its synergy effect [19]. Zhang et al. also proposed a straightforward and environmentally benign method for creating core-shell hybrid electrocatalyst $Fe/Fe_3C@NC-G$ via carbonization and hydrothermal treatment. In an alkaline electrolyte, the catalysts demonstrated a strong catalytic ORR ability. Furthermore, the catalyst also outperforms the currently available Pt/C catalyst in alkaline media in terms of durability and methanol tolerance. The ORR activity of hybrids is thought to be improved by the synergy between the nitrogen-doped graphite shell anchored on graphene and the Fe/Fe_3C core, as well as by the mesoporous structure. Furthermore, the carbon shell may guarantee the high stability of electrocatalyst and stop the deterioration of ORR activity [20].

Polymer Metal Hybrid Electrocatalyst: Metal polymer hybrid electrocatalysts find its application especially for energy conversion like fuel cells, water splitting (electrolysis), and CO_2 reduction. To enhance the conductivity, stability, and ion transport characteristics of hybrid electrocatalysts, polymers such as polypyrrole (PPy), polyaniline (PANI), and polythiophene can be added. They can contribute to the overall catalytic activity and serve as binders or support materials. One of the most promising methods to use ample CO_2 feedstocks in ambient circumstances is the electrochemical conversion of CO_2 into value-added products. Although metal-based heterogeneous catalysts frequently show significant activity for the CO_2 reduction reaction (CO_2RR), it is still difficult to achieve high activity, selectivity, and stability all at once. The hybrid approach has proven to be a successful way to address the aforementioned issue. The ability to regulate the active site motifs' surroundings is one of the benefits of metal–polymer hybrids. By using a simple modification technique, metal/polymer hybrids have so far been effectively used as electrocatalysts for CO_2RR. In their research, Jia et al. showed that altering the polymer on the metal surface was a useful tactic for enhancing CO_2RR performance by fabricating 3D hierarchical M/polymer-CP electrodes for CO_2 electro-reduction via in situ electrodeposition [21].

Electrocatalysts and Advanced Materials for Sustainable Energy Storage　　Materials Research Forum LLC
Materials Research Foundations 182 (2025) 57-69　　　　https://doi.org/10.21741/9781644903797-5

Srabanti et.al in their review article provide an overview on the use of conducting polymer-based nanohybrids as active electrode materials or catalyst supports for PEM and MFC applications. Since conducting polymers have special qualities like high surface area, improved electron delocalization from the CP to the hybrid metal, and high electronic conductivity because of the conjugated backbone, the CPNH is a desirable material to investigate its physicochemical characteristics and expand its use in various PEMFC types [22]. A N- and Si-doped polythiophene nanocomposite (PTh-NSi) is presented by Bian et al. as a biocompatible non-metal HER electrocatalyst for use in a hybrid microbial-inorganic catalysis system offers a polythiophene nanocomposite (PTh-NSi) doped with N and Si as a biocompatible nonmetal HER electrocatalyst for application in a hybrid microbial-inorganic catalysis system. The electrocatalyst PTh-NSi obtained a 10 mA/cm^2 current density in neutral PBS with a low overpotential of -0.395 V, demonstrating outstanding HER activity [23]. The stability, conductivity, and ionic transport of hybrid electrocatalysts can all be further enhanced by the addition of polymers. It is common practice to integrate metal nanoparticles and carbon supports into polymers such as polypyrrole or polyaniline. By limiting sintering and promoting better dispersion of the metal nanoparticles, the polymer matrix can stabilize the catalyst and increase its long-term stability.

Carbon based Hybrid Electrocatalyst: Carbon based hybrid electrocatalysts are advanced materials designed for efficient electrochemical reactions. These are primarily composed of carbon materials, such as graphene, carbon nanotubes, or porous carbon. They are often doped with heteroatoms like nitrogen, sulphur, or boron to enhance their catalytic activity. Carbon-based electrocatalysts are metal-free and are valued for their high conductivity, chemical stability, and environmental friendliness. They are typically used in applications where cost-effectiveness and sustainability are priorities, such as in oxygen reduction reactions (ORR) in fuel cells. Graphene, carbon nanotubes (CNTs), carbon black, and mesoporous carbon are widely used due to their high surface area, electrical conductivity, and stability. Carbon-supported metals such as Pt/C and Pd/C are commonly used in fuel cells and batteries for reactions like ORR and HER. Fan et al., demonstrated that Silica Carbide, Graphene diamond clusters, and Graphene Nanoribbons self-assemble into an integrated nanostructure in this SiC-GD@GNRs composite, resulting in outstanding HER activity with a tiny onset overpotential of 8 mV. The catalyst may operate steadily for a long time and drive a high cathodic current at a low overpotential. Advanced metal-free electrocatalysts can be easily generated using this material synthesis method, and they can be easily integrated with a variety of water-splitting devices [24].

Liang et al. investigated ORR, OER, and HER, several inorganic/nanocarbon hybrid materials have been created as effective electrocatalysts. In order to add functional groups for the nucleation, growth, and anchoring of inorganic materials while retaining a comparatively high electrical conductivity, nanocarbon materials, such as graphene and carbon nanotubes, are appropriately oxidized. Strong chemical and electrical coupling between the inorganic nanomaterials and the functionalized nanocarbon is produced as a result of the direct nucleation and growth of the inorganic nanomaterials on the nanocarbon. This improved charge transport through interfaces allowed the hybrid materials to have high electrochemical performance [25]. Chang Yu et al. synthesized a hydrothermal method for creating an all-carbon hybrid electrocatalyst made up of N-doped carbon atoms as the catalytic sites and graphene as a support. In an alkaline medium, the material shows superior stability, excellent selectivity, and competitive electrocatalytic activity when compared to commercial Pt-based catalysts. As a result, it exhibits great potential as an inexpensive and effective metal-free catalyst for ORR. This kind of hybrid will offer the chance to create a variety of effective, metal-free ORR catalysts, which are necessary for real-world fuel cell applications [26]. Table 1 gives a comparative analysis of various hybrid electrocatalysts [27-31].

Table 1. Comparative analysis of different hybrid electrocatalysts

Composite Type	Catalyst	Electrolyte	Deposition Technique	Current density (mA cm^{-2})	Tafel slope (mV dec^{-1})	Ref.
Metal carbon	Fe-doped CoP nanosheet/ carbon nanotube	1 M KOH	Doping	10	24	[27]
Metal Oxide	CoO/Fe$_3$O$_4$	1 M KOH	Annealing (thermal reduction in Ar gas flow)	20	73	[28]
Core-shell nanostructure	Pd loaded nano spindle	0.1M KOH	Drop casting	0.1	93.8	[29]
Polymer metal	poly(a-terthiophene) -Pt nanoparticle hybrid material (polyTT-Pt)	0.1 M KCl	in situ growth	10	31	[30]
Carbon based	N-CNTs	0.1M KOH	CVD	2	54	[31]

Advancements and Innovations

At the forefront of sustainable energy technology development are hybrid electrocatalysts, especially for uses such as water splitting (electrolysis), metal-air batteries, and fuel cells. These materials provide synergistic effects that improve performance by combining several catalytic components, including metals, metal oxides, carbon-based compounds, and conductive polymers. Recently, many other types of hybrid electrocatalysts have been explored which can provide better catalytic efficiency. Hybrid catalysts where single metal atoms (e.g., Fe, Co, Ni) are embedded in nitrogen-doped carbon frameworks. These offer high activity and selectivity for oxygen reduction reaction (ORR) and oxygen evolution reaction (OER) with much lower cost than platinum-group metals. Combining metal Nitrogen-Carbon (M–N–C) materials with metal nanoparticles or carbon nanotubes to create hierarchical structures improves conductivity and active site exposure. Recently, Carbonized MOFs forming porous carbon matrices doped with metal/metal oxide are also being extensively used for example, Co/N-doped carbon from ZIF-67 for OER and ORR. Conductive 2D materials like Ti$_3$C$_2$Tx, combined with metals or oxides, are also used for improving HER/OER activity. Yang et.al has used an amorphous nanoporous PdCuNi-S catalyst for highly effective HER. The property of the material is attributed to the bicontinuous nanoporous structure, enhanced charge transfer (Figure 2). Hybridization alters electronic structures, improving charge transfer kinetics. Interface-induced lattice strain modifies binding energies for intermediates. Combining materials with complementary catalytic functions for dual reactions (e.g., OER + HER).

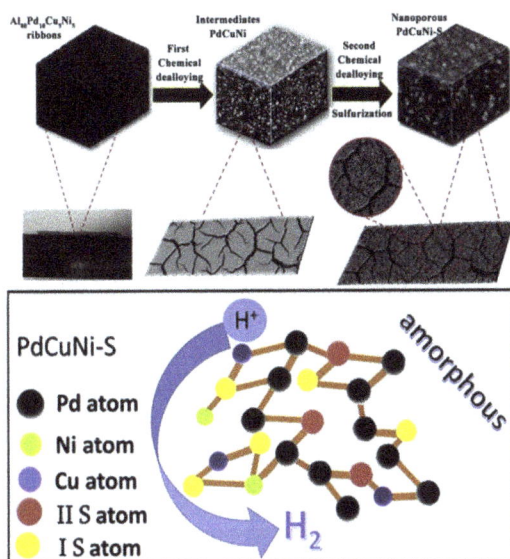

Figure 2. Synthesis method for nanoporous PdCuNi-S catalyst with HER evolution mechanism (Reproduced with permission from Ref. [32], Copyright 2019, Elsevier).

Challenges and Future Directions

Many hybrid catalysts face challenges related to long-term stability, especially under extreme conditions like high temperatures or acidic media. Many hybrid catalysts are designed and tested on a small scale, scaling up these materials for industrial applications maintaining their performance is a challenge. For maximizing performance and minimizing cost, finding the right balance between the components (e.g., metal loading, metal type, carbon support) is essential. Creating hybrid electrocatalysts by combining several active sites or functions into a single composite or substance is a promising method in which complementary functions can be demonstrated by hybrid systems that combine metal-based catalysts with semiconductor or carbon-based materials. For instance, combining earth-abundant materials like transition metal oxides or nitrides with noble metals like platinum may increase stability, lower costs, and improve overall catalytic performance. The logical design of hybrid materials to maximize the reactivity and electrical characteristics of the constituent parts will be the main focus. More effective catalysts for the CO_2 reduction, oxygen evolution reaction, and hydrogen evolution reaction (HER) may result from this. A better understanding of how various materials interact at the interface in hybrid electrocatalysts can improve their efficiency. The synergy between various materials at their interfaces is frequently the basis for hybrid electrocatalysts. To maximize their catalytic performance, it will be essential to comprehend how charge transfer, atomic rearrangements, and electronic coupling take place at these interfaces. Researchers will be able to alter the interface chemistry through sophisticated methods like in-situ analysis and computational modelling, increasing the stability and activity of hybrid catalysts. Addressing the hybrid electrocatalysts long-term stability and robustness in challenging reaction circumstances is an expected advancement in the field of hybrid electrocatalysts. Leaching, sintering, and corrosion are some of the common causes of hybrid electrocatalyst depreciation or loss of activity. Future studies will try to create stronger hybrid materials that can withstand these types of degradation, either by adding stabilizing chemicals or protective coatings or by making the materials more resilient naturally. The commercialization of hybrid catalysts in energy systems like fuel cells, batteries, and electrolyzers will depend on the creation of robust and long-lasting catalysts. Generate hybrid electrocatalysts using biomimetic principles that are modelled after natural systems. In reactions like CO_2 reduction or N_2 fixation, hybrid electrocatalysts could be created to have great selectivity and efficiency by imitating natural catalysts (such as those in photosynthesis or nitrogen fixation). These biohybrids might blend synthetic materials (like metals or semiconductors) with biological elements (like proteins or enzymes).

In energy and environmental applications, the combination of synthetic materials and biological catalysts may create new opportunities for highly sustainable and selective reactions. Developing catalysts using more widely available, cost-effective materials instead of relying on rare and costly materials like iridium and platinum. Promising substitutes are provided by transition metal-based catalysts, especially those that combine conductive carbon-based materials (such as graphene) or polymers with nickel, cobalt, iron, or manganese. Performance may be enhanced by hybridizing these with high-surface-area or nanostructured materials. In order to lower the financial hurdles to broad adoption, a focus will be placed on creating high-performance hybrid catalysts from inexpensive, plentiful materials that can sustain high activity and stability over time. The creation of catalysts with exact atomic configurations may open up new reaction pathways and greatly improve electrocatalytic activity. In order to contribute to a more resilient and sustainable energy grid, efforts will be directed on developing hybrid electrocatalysts that can function well in variable environments, like those present in intermittent renewable energy sources. High-performance materials will be developed more quickly, attributable to the incorporation of data-driven techniques, which will make it possible to screen potential hybrid catalysts more effectively and optimize synthesis procedures.

Conclusion

Hybrid electrocatalysts represent a promising solution for improving the efficiency and sustainability of electrochemical processes like fuel cells, water splitting, and energy storage. By strategically combining different materials, researchers can exploit the individual strengths of each component to achieve enhanced catalytic activity, better stability, and lower costs. The ongoing challenge is to optimize these hybrids, ensuring they meet the requirements for large-scale industrial applications while maintaining high performance. Hybrid electrocatalysts have a bright future beyond them, with several viable avenues to improve sustainability, performance, and affordability. Hybrid electrocatalysts have the potential to completely transform energy conversion and storage technologies as long as we keep combining different materials, improving interface interactions, and utilizing state-of-the-art tools like artificial intelligence. Hybrid electrocatalysts may be essential in accelerating the shift to sustainable energy by tackling important issues, including stability, cost, and scalability.

References

[1] Crabtree, George W., and Mildred S. Dresselhaus. "The hydrogen fuel alternative." *Mrs Bulletin* 33, no. 4 (2008): 421-428. https://doi.org/10.1557/mrs2008.84

[2] Chen, Houmao, Xianyou Luo, Shaopeng Huang, Feng Yu, De Li, and Yong Chen. "Phosphorus-doped activated carbon as a platinum-based catalyst support for electrocatalytic hydrogen evolution reaction." *Journal of Electroanalytical Chemistry* 948 (2023): 117820. https://doi.org/10.1016/j.jelechem.2023.117820

[3] Wu, Jianbo, and Hong Yang. "Platinum-based oxygen reduction electrocatalysts." *Accounts of chemical research* 46, no. 8 (2013): 1848-1857. https://doi.org/10.1021/ar300359w

[4] Wang, Kun, Yi Wang, Yexiang Tong, Zhengwei Hao Pan, and Shuqin Song. "A robust versatile hybrid electrocatalyst for the oxygen reduction reaction." *ACS Applied Materials & Interfaces* 8, no. 43 (2016): 29356-29364. https://doi.org/10.1021/acsami.6b03751

[5] Yang, Fulin, Pingping Zhao, Xing Hua, Wei Luo, Gongzhen Cheng, Wei Xing, and Shengli Chen. "A cobalt-based hybrid electrocatalyst derived from a carbon nanotube inserted metal–organic framework for efficient water-splitting." *Journal of Materials Chemistry A* 4, no. 41 (2016): 16057-16063. https://doi.org/10.1039/C6TA05829A

[6] Perivoliotis, D. K., and N. Tagmatarchis. "Recent advancements in metal-based hybrid electrocatalysts supported on graphene and related 2D materials for the oxygen reduction reaction." *Carbon* 118 (2017): 493-510. https://doi.org/10.1016/j.carbon.2017.03.073

[7] Bai, Song, and Yujie Xiong. "Some recent developments in surface and interface design for photocatalytic and electrocatalytic hybrid structures." *Chemical Communications* 51, no. 51 (2015): 10261-10271. https://doi.org/10.1039/C5CC02704G

[8] Yang, Jinhui, Donge Wang, Hongxian Han, and C. A. N. Li. "Roles of cocatalysts in photocatalysis and photoelectrocatalysis." *Accounts of chemical research* 46, no. 8 (2013): 1900-1909. https://doi.org/10.1021/ar300227e

[9] Long, Ran, Shan Zhou, Benjamin J. Wiley, and Yujie Xiong. "Oxidative etching for controlled synthesis of metal nanocrystals: atomic addition and subtraction." *Chemical Society Reviews* 43, no. 17 (2014): 6288-6310. https://doi.org/10.1039/C4CS00136B

[10] Bai, Song, Xijun Wang, Canyu Hu, Maolin Xie, Jun Jiang, and Yujie Xiong. "Two-dimensional gC 3 N 4: an ideal platform for examining facet selectivity of metal co-catalysts in photocatalysis." *Chemical communications* 50, no. 46 (2014): 6094-6097. https://doi.org/10.1039/C4CC00745J

[11] Xiao, Fei, Gui-Liang Xu, Cheng-Jun Sun, Inhui Hwang, Mingjie Xu, Hsi-wen Wu, Zidong Wei, Xiaoqing Pan, Khalil Amine, and Minhua Shao. "Durable hybrid electrocatalysts for proton exchange membrane fuel cells." *Nano Energy* 77 (2020): 105192. https://doi.org/10.1016/j.nanoen.2020.105192

[12] Jha, Neetu, R. Imran Jafri, N. Rajalakshmi, and S. Ramaprabhu. "Graphene-multi walled carbon nanotube hybrid electrocatalyst support material for direct methanol fuel cell." *International journal of hydrogen energy* 36, no. 12 (2011): 7284-7290. https://doi.org/10.1016/j.ijhydene.2011.03.008

[13] Arici, Ece, Begum Yarar Kaplan, Ahmet Musap Mert, Selmiye Alkan Gursel, and Solen Kinayyigit. "An effective electrocatalyst based on platinum nanoparticles supported with graphene nanoplatelets and carbon black hybrid for PEM fuel cells." *International Journal of Hydrogen Energy* 44, no. 27 (2019): 14175-14183. https://doi.org/10.1016/j.ijhydene.2018.11.210

[14] Zhu, Yinlong, Qian Lin, Yijun Zhong, Hassan A. Tahini, Zongping Shao, and Huanting Wang. "Metal oxide-based materials as an emerging family of hydrogen evolution electrocatalysts." *Energy & Environmental Science* 13, no. 10 (2020): 3361-3392. https://doi.org/10.1039/D0EE02485F

[15] Pratama, Dwi Sakti Aldianto, Andi Haryanto, and Chan Woo Lee. "Heterostructured mixed metal oxide electrocatalyst for the hydrogen evolution reaction." *Frontiers in Chemistry* 11 (2023): 1141361.https://doi.org/10.3389/fchem.2023.1141361

[16] Wei, Guijuan, Chunyang Wang, Xixia Zhao, Shoujuan Wang, and Fangong Kong. "Plasma-assisted synthesis of Ni4Mo/MoO2@ carbon nanotubes with multiphase-interface for high-performance overall water splitting electrocatalysis." *Journal of Alloys and Compounds* 939 (2023): 168755. https://doi.org/10.1016/j.jallcom.2023.168755

[17] Singh, K. P., Shin, C. H., Lee, H. Y., Razmjooei, F., Sinhamahapatra, A., Kang, J., et al. (2020). TiO2/ZrO2 nanoparticle composites for electrochemical hydrogen evolution. ACS Appl. Nano Mat. 3, 3634–3645. https://doi.org/10.1021/acsanm.0c00346

[18] Guo, Shaojun, Youxing Fang, Shaojun Dong, and Erkang Wang. "High-efficiency and low-cost hybrid nanomaterial as enhancing electrocatalyst: spongelike Au/Pt core/shell nanomaterial with hollow cavity." *The Journal of Physical Chemistry C* 111, no. 45 (2007): 17104-17109. https://doi.org/10.1021/jp075625z

[19] Magro, Massimiliano, Davide Baratella, Gabriella Salviulo, Katerina Polakova, Giorgio Zoppellaro, Jiri Tucek, Josef Kaslik, Radek Zboril, and Fabio Vianello. "Core–shell hybrid nanomaterial based on prussian blue and surface active maghemite nanoparticles as stable electrocatalyst." *Biosensors and Bioelectronics* 52 (2014): 159-165. https://doi.org/10.1016/j.bios.2013.08.052

[20] Zhang, Yating, Peng Wang, Juan Yang, Keke Li, Xueying Long, Meng Li, Kaibo Zhang, and Jieshan Qiu. "Fabrication of core-shell nanohybrid derived from iron-based metal-organic framework grappled on nitrogen-doped graphene for oxygen reduction reaction." *Chemical Engineering Journal* 401 (2020): 126001. https://doi.org/10.1016/j.cej.2020.126001

[21] Jia, Shuaiqiang, Qinggong Zhu, Mengen Chu, Shitao Han, Ruting Feng, Jianxin Zhai, Wei Xia, Mingyuan He, Haihong Wu, and Buxing Han. "Hierarchical metal–polymer hybrids for enhanced CO2 electroreduction." *Angewandte chemie international edition* 60, no. 19 (2021): 10977-10982. https://doi.org/10.1002/anie.202102193

[22] Ghosh, Srabanti, Suparna Das, and Marta EG Mosquera. "Conducting polymer-based nanohybrids for fuel cell application." *Polymers* 12, no. 12 (2020): 2993https://doi.org/10.3390/polym12122993

[23] Bian, Xianghai, Yang Ye, Sulin Ni, Bin Yang, Yang Hou, Lecheng Lei, Min Yao, and Zhongjian Li. "Polythiophene-Based Nonmetal Electrocatalyst with Biocompatibility to Boost Efficient CO2 Conversion." *Chem & Bio Engineering* (2025). https://doi.org/10.1021/cbe.4c00156

[24] Fan, Xiujun, Zhiwei Peng, Juanjuan Wang, Ruquan Ye, Haiqing Zhou, and Xia Guo. "Carbon-Based Composite as an Efficient and Stable Metal-Free Electrocatalyst." *Advanced Functional Materials* 26, no. 21 (2016): 3621-3629. https://doi.org/10.1002/adfm.201600076

[25] Liang, Yongye, Yanguang Li, Hailiang Wang, and Hongjie Dai. "Strongly coupled inorganic/nanocarbon hybrid materials for advanced electrocatalysis." *Journal of the American Chemical Society* 135, no. 6 (2013): 2013-2036. https://doi.org/10.1021/ja3089923

[26] Hu, Chao, Chang Yu, Mingyu Li, Xiuna Wang, Qiang Dong, Gang Wang, and Jieshan Qiu. "Nitrogen-doped carbon dots decorated on graphene: a novel all-carbon hybrid electrocatalyst for enhanced oxygen reduction reaction." *Chemical communications* 51, no. 16 (2015): 3419-3422. https://doi.org/10.1039/C4CC08735F

[27] Zhang, Xing, Xiao Zhang, Haomin Xu, Zishan Wu, Hailiang Wang, and Yongye Liang. "Iron-doped cobalt monophosphide nanosheet/carbon nanotube hybrids as active and stable electrocatalysts for water splitting." *Advanced Functional Materials* 27, no. 24 (2017): 1606635. https://doi.org/10.1002/adfm.201606635

[28] Adamson, William, Xin Bo, Yibing Li, Bryan HR Suryanto, Xianjue Chen, and Chuan Zhao. "Co-Fe binary metal oxide electrocatalyst with synergistic interface structures for efficient overall water splitting." *Catalysis Today* 351 (2020): 44-49. https://doi.org/10.1016/j.cattod.2019.01.060

[29] Yusuf, Mohammad, Muthuchamy Nallal, Ki Min Nam, Sehwan Song, Sungkyun Park, and Kang Hyun Park. "Palladium-loaded core-shell nanospindles as potential alternative electrocatalyst for oxygen reduction reaction." *Electrochimica Acta* 325 (2019): 134938. https://doi.org/10.1016/j.electacta.2019.134938

[30] Chakrabartty, Sukanta, Chinnakonda S. Gopinath, and C. Retna Raj. "Polymer-based hybrid catalyst of low Pt content for electrochemical hydrogen evolution." *International Journal of Hydrogen Energy* 42, no. 36 (2017): 22821-22829. https://doi.org/10.1016/j.ijhydene.2017.07.152

[31] Li, Xuelian, Jingwen Zhou, Junxiang Zhang, Matthew Li, Xuanxuan Bi, Tongchao Liu, Tao He et al. "Bamboo-like nitrogen-doped carbon nanotube forests as durable metal-free catalysts for self-powered flexible Li–CO2 batteries." *Advanced Materials* 31, no. 39 (2019): 1903852. https://doi.org/10.1002/adma.201903852

[32] Yang, Xinxin, Wence Xu, Shuo Cao, Shengli Zhu, Yanqin Liang, Zhenduo Cui, Xianjin Yang et al. "An amorphous nanoporous PdCuNi-S hybrid electrocatalyst for highly efficient hydrogen production." *Applied Catalysis B: Environmental* 246 (2019): 156-165. https://doi.org/10.1016/j.apcatb.2019.01.030

Electrocatalysts and Advanced Materials for Sustainable Energy Storage   Materials Research Forum LLC
Materials Research Foundations 182 (2025) 70-86   https://doi.org/10.21741/9781644903797-6

Chapter 6

Carbon Nanotube based Catalysts for Energy Storage Studies: Synthesis and Electrochemical Analysis

LAVEENA Mariet Veigas[1,2,a], ANISHA Guha[2,b], HERI SEPTYA Kusuma[3,c],
MOTHI Krishna Mohan[2,d*]

[1]St. Joseph's University, 36 Lalbagh Road, Bengaluru 560027, Karnataka, India

2Department of Sciences and Humanities, School of Engineering and Technology, Christ
University, Bangalore-560074, India

[3]Department of Chemical Engineering, Faculty of Industrial Technology, Universitas

Pembangunan Nasional "Veteran" Yogyakarta, Indonesia

[a]lmveigas@gmail.com, [b]anisha.guha@res.christuniversity.in, [c]heriseptyakusuma@gmail.com,
[d]mothikrishna.mohan@christuniversity.in

Abstract

Energy storage technologies are pivotal in transitioning to sustainable energy solutions. Catalysts improve the energy storage efficiency, enhancing reaction kinetics and enabling advanced electrochemical processes. Carbon nanotubes (CNTs) are noteworthy catalyst materials due to their exceptional electrical conductivity, large surface area, and unique structural properties. This chapter provides a thorough overview of Carbon nanotube-based catalysts for energy storage applications, covering their synthesis, functionalization, electrochemical characterization, and practical implementations. The synthesis of carbon nanotube-based catalysts is examined, focusing on methods like chemical vapor deposition (CVD), arc discharge, laser ablation, and plasma-enhanced synthesis. Functionalization techniques, including surface modification, metal/non-metal doping, composite formation with metal oxides and conducting polymers, are examined to understand their role in enhancing catalytic performance. Electrochemical analysis is crucial in evaluating the efficiency of carbon nanotube-based catalysts. This chapter deals with essential techniques such as cyclic voltammetry (CV), electrochemical impedance spectroscopy (EIS), chronoamperometry, and Tafel analysis, emphasizing their role in understanding charge storage mechanisms. The application of carbon nanotube-based catalysts in supercapacitors, lithium-ion/sodium-ion batteries, fuel cells, and hydrogen evolution reactions is explored, discussing their impact on energy storage performance. Despite their advantages, challenges in areas such as large-scale synthesis, stability, and cost remain a point of concern. The chapter concludes by discussing future directions and the potential impact of CNT-based catalysts in advancing next-generation energy storage technologies.

Keywords

Carbon Nanotubes, Electrocatalyst, Energy Storage, Electrochemical Analysis, Synthesis

Contents

Introduction

The increasing focus on sustainable and renewable energy sources has heightened the demand for advanced energy storage technologies. Effective storage solutions are essential for managing the intermittent nature of renewable sources such as solar and wind. They provide a consistent power output and improve overall energy efficiency. Technologies such as batteries, supercapacitors, and fuel cells have become central to addressing these challenges, each with specific advantages concerning energy capacity, power density, and durability. However, their performance largely depends on various catalysts and materials integrated into their design, especially at the electrode interfaces. Catalysts play a crucial role in energy storage systems by accelerating electrochemical reactions, lowering energy barriers, and improving electron transfer rates. Their presence can significantly increase the efficiency and lifespan of energy storage devices, making the design of robust and efficient catalysts a key area of research. Despite numerous advancements, finding catalysts that combine high activity, stability, and cost-effectiveness remains a significant challenge. Carbon nanotubes (CNTs) have emerged as up-and-coming materials because of their unique structural, electrical, and mechanical properties that distinguish them from conventional catalysts. CNTs exhibit exceptional electrical conductivity, large surface area for catalyst dispersion, tensile strength, and chemical resilience, all contributing to improved performance. Moreover, CNTs can be synthesized in various forms, like single and multi-walled nanotubes, offering versatility in customizing their properties for particular applications. Functionalization and doping techniques further expand their potential by introducing active sites that boost catalytic activity.

This chapter intends to provide a comprehensive run-through of CNT-based catalysts within the context of energy storage technologies. It covers fundamental aspects such as CNT synthesis methods, like chemical vapor deposition, arc discharge, and laser ablation, alongside strategies for functionalizing and modifying CNTs to enhance their catalytic properties. Additionally, the chapter discusses electrochemical characterization techniques, such as cyclic voltammetry and impedance spectroscopy, which are essential for evaluating catalyst performance. This chapter also highlights how CNT-based catalysts contribute to the advancement of supercapacitors, lithium-ion, sodium-ion batteries, and fuel cells, emphasizing their role in improving charge storage, conductivity, and catalytic efficiency. It also addresses challenges related to scalability, stability, and cost, concluding with insights into future research directions for leveraging CNTs in next-generation energy storage systems.

Synthesis of Carbon Nanotubes-Based Catalysts

Design and synthesis of efficient, sustainable, and cost-effective materials represent some of the most effective methods to enhance energy utilisation effectiveness, alleviating significant energy and environmental issues. There are several effective methods for synthesizing carbon nanotubes, including arc discharge, laser ablation, and chemical vapor deposition (CVD). Among these methods, CVD is the most promising and low-cost production method [1,2].

Chemical Vapour Deposition (CVD): In CVD-based CNT production, gases like CH_4 and CO are decomposed by metal catalysts in a high-temperature chamber at pressures of 1 atm or lower. The carbon atoms separate from the gas as it enters the chamber, nucleating on the metal catalysts and subsequently assembling into cylindrical CNTs. The resulting CNT material is then removed from the furnace after cooling to prevent damage to the tubes from oxidation [3].

Arc Discharge (AD): This approach employs a graphite electrode with a gap of a few millimeters to generate carbon plasma. It occurs in a sealed chamber maintained at or below 1 atm pressure, where vapours condense on a metal catalyst. As the plasma approaches the cathode, it releases a stream of particles. This method is typically utilised in the production of C60, where various graphitic materials may be present in the final product, requiring post-production purification and separation of CNTs [4].

Laser Ablation (LA): This technique uses a laser to vaporise graphite in a high-temperature chamber. Typically, the graphitic material is combined with catalyst metals, which facilitates the formation of nanotubes as the vaporised material condenses on the chamber's cooled surfaces [5].

Every synthesis method for carbon nanotubes (CNTs) involves specific process variables influencing their development and properties. These include gas flow rates, pressure, reaction time, growth temperature, catalyst size and concentration, nature of substrates, carrier gases, and carbon precursors. It is essential to optimise these parameters to achieve a pure, well-structured CNT with a higher yield [6].

Functionalization and Modification of CNTs

To use CNTs effectively as catalysts or supports, surface modification is critical for improving their chemical reactivity, dispersion in solvents, and binding with active catalytic materials.

Surface Functionalisation for Catalytic Activity: Surface functionalisation enhances the chemical compatibility and dispersion of CNTs, introduces active sites, and increases surface defects. It is typically achieved in the following ways:

➢ *Oxidative treatments:* Acid treatments using HNO_3 or H_2SO_4 that introduce carboxyl (-COOH), hydroxyl (-OH), or carbonyl (=O) groups, enhancing wettability and reactivity,

➢ *Plasma treatment:* Oxygen or ammonia plasma introduces oxygen or nitrogen-containing groups without much altering the CNT structure,

➢ *Silane coupling:* Introduces linker molecules that anchor catalytic species [7,8].

Doping with Metals and Non-Metals: Doping introduces heteroatoms into the carbon lattice of CNTs, which significantly alters their electronic and catalytic properties.

➢ *Metal doping:* Metals such as Fe, Co, Ni, and Pt enhance electrocatalytic activity in reactions like oxygen reduction (ORR) and hydrogen evolution (HER) by depositing metal nanoparticles on CNT surfaces.

➢ *Non-metal doping (e.g., N, B, P, S):* Nitrogen-doped carbon nanotubes (N-CNTs) enhance catalytic activity for the oxygen reduction reaction (ORR) and hydrogen evolution reaction (HER) through their lone-pair electrons. Other dopants like boron or sulphur create strain or electronic effects that alter the reactivity [9,10].

Composite Formation with Metal Oxides and Conducting Polymers: Combining CNTs with other electroactive materials enhances their catalytic performance and synergistically improves charge transport, mechanical stability, and electrochemical reactivity.

➢ *CNT/metal oxide composites*: Incorporating materials such as MnO_2, Fe_2O_3, NiO, and Co_3O_4 enhances pseudo capacitance and redox kinetics, which is helpful in batteries and supercapacitors.

➢ *CNT/conducting polymer composites:* Integrating polymers like polyaniline (PANI) or polypyrrole (PPy) enhances surface area and flexibility, rendering the polymer ideal for wearable and flexible energy devices.

➢ *CNT–graphene hybrids:* These materials merge the conductivity of CNTs with the high surface area and flexibility of graphene, thereby enhancing ion transport and mechanical durability [11,12].

The synthesis and modification of CNTs play a pivotal role in dramatically enhancing their effectiveness as catalysts in energy storage applications. Chemical Vapour Deposition (CVD) is still the highly effective method for large-scale CNT manufacture, while chemical functionalisation, doping, and composite creation adjust their properties for specific catalytic applications. Ongoing advancements in scalable synthesis and molecular-level engineering are anticipated to elevate the potential of CNT-based catalysts in next-generation electrochemical systems.

Electrochemical Analysis of CNT-Based Catalysts

Electrochemical analysis is an essential technique for evaluating the efficiency, activity, and stability of catalysts, particularly those of carbon nanotubes (CNTs). The foundation of this analysis is the electrochemical workstation. This instrument enables the control and measurement of electrical parameters such as current, voltage, and impedance in a well-defined electrochemical system. A typical workstation comprises a potentiostat/galvanostat, electrodes (working, counter, and reference), and data acquisition and analysis [13]software. For CNT-based catalysts, electrochemical workstations facilitate the investigation of electron transfer properties and surface reactivity. Given the high conductivity and large surface area of CNTs, these catalysts are often tested under varying electrochemical conditions to understand their behaviour under different loadings, functionalization, and composite forms [14].

Importance of Electrochemical Analysis in Catalyst Evaluation

Electrochemical analysis is indispensable in catalyst evaluation because it provides direct insight into the kinetic and thermodynamic properties of the catalytic surface. For CNT-based materials, which often serve as support structures or active catalysts themselves, the analysis reveals how surface modifications, doping, and structural integrity influence the overall catalytic performance [15]. Additionally, electrochemical techniques help determine the durability and degradation mechanisms of CNT-based catalysts under prolonged use. Unlike purely structural or spectroscopic evaluations, electrochemical methods offer real-time monitoring of changes in activity, making them highly effective for iterative optimization of catalyst design [16].

Common Electrochemical Techniques for CNT-Based Catalysts

A variety of electrochemical techniques are employed to investigate CNT-based catalysts. Each technique provides unique information about different aspects of catalyst behaviour, from redox activity to charge transfer kinetics and diffusion characteristics [17].

Cyclic Voltammetry (CV): Cyclic voltammetry is a common electrochemical technique used to analyze catalysts based on carbon nanotubes (CNTs). It includes linearly sweeping the electrode potential over time and recording the current response. The resulting cyclic voltammogram provides information on redox processes, electrochemical reversibility, and capacitive behaviour [18]. For CNT-based systems, CV is beneficial in detecting shifts in peak potentials and variations in peak current intensity, which are indicative of changes in catalytic activity or surface modification. CNTs functionalized with heteroatoms or nanoparticles often exhibit altered voltametric profiles, which can be quantitatively analysed to assess enhancement or degradation in performance [19,20].

Electrochemical Impedance Spectroscopy (EIS): Electrochemical impedance spectroscopy is a technique that measures the impedance of a system across a range of alternating current (AC) frequencies. It provides insight into charge transfer resistance and double-layer capacitance at the electrode-electrolyte interface [21]. In the context of CNT-based catalysts, EIS is particularly valuable for assessing the electrical conductivity and interface characteristics [22]. The Nyquist and Bode plots generated through EIS can reveal resistive and capacitive elements in the system, allowing researchers to model equivalent circuits that describe the electrochemical behaviour of CNT composites. A lower charge transfer resistance typically signifies improved catalytic efficiency [23,24].

Chronoamperometry and Chronopotentiometry: Chronoamperometry (CA) and chronopotentiometry (CP) are time-dependent techniques used to evaluate the stability and durability of electrochemical systems. In CA, a fixed potential is applied, and the resulting current is measured over time, whereas in CP, a fixed current is applied, and the voltage response is monitored [25]. These techniques, when applied to CNT-based catalysts, enhance our understanding of the material's long-term electrochemical stability under constant operational conditions. Any decline in current (in CA) or increase in potential (in CP) over time may suggest degradation, dissolution, or passivation of the active surface. These results are especially critical for catalysts subjected to high operational stress or harsh chemical environments [26].

Tafel Analysis and Polarization Curves: Tafel analysis involves plotting the logarithm of current density against the overpotential to determine kinetic parameters like exchange current density and Tafel slope. Polarization curves, which plot current density versus potential, are also used to assess the electrocatalytic activity [27]. For CNT-based materials, Tafel analysis can help quantify the intrinsic catalytic activity by highlighting the rate-determining step of the electrochemical reaction. A lower Tafel slope typically suggests faster kinetics and a more efficient catalyst. These curves also help compare performance between various CNT modifications, enabling a clearer understanding of the role played by surface chemistry and structural morphology [28,29].

Electrochemical Performance in Energy Storage Applications

Though not delving into application benefits, it is essential to discuss how CNT-based catalysts perform under specific electrochemical tests tailored for energy storage systems. These assessments provide benchmark data for understanding their suitability and limitations in different energy environments [30].

Performance in Supercapacitors: Electrochemical tests for CNT-based catalysts in supercapacitors often involve repetitive CV cycles and galvanostatic charge-discharge measurements. Here, the focus is on evaluating the capacitance retention, charge-discharge symmetry, and coulombic efficiency [31]. CNTs exhibit pseudocapacitive behaviour depending on their functionalization and composite formation. Electrochemical analysis in this context plays a crucial role in distinguishing whether the storage of charge is generated through the formation of electric double layers or redox reactions. This understanding is essential for advancing our knowledge and improving the effectiveness of energy storage systems. The shape of the CV curve is typically rectangular for ideal capacitors, and the slope of the charge-discharge curves serves as an indicator of capacitive efficiency and material stability [32].

(i) Role in Lithium-Ion and Sodium-Ion Batteries: CV and EIS are commonly employed to probe the electrode kinetics and interfacial behaviour of CNT-based catalysts for battery analysis. The presence of distinct redox peaks in CV curves indicates lithium or sodium intercalation and de-intercalation processes [33]. EIS measurements offer essential insights into charge transfer resistance as well as solid electrolyte interface (SEI) resistance. The reduction in impedance after cycling may suggest improved ion transport, while an increase could signal degradation. These analyses help identify the phase transitions and mechanical stability of CNT-based electrodes under repeated cycling [34]. Electrochemical techniques enhance the growth of vertically aligned CNTs as shown in Figure 1, improving the electrode performance in supercapacitors and other electrochemical devices.

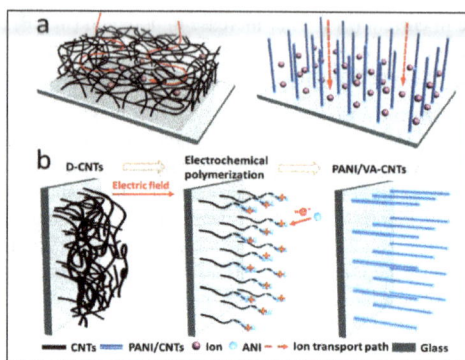

Figure 1. Schematic representation of Electrochemical-induced Vertical-aligned Carbon nanotubes and polyaniline nanocomposite electrode structure (Reproduced with permission from Ref. [35], Copyright 2017, Springer Nature).

(ii) CNTs in Fuel Cells and Electrocatalysis: In the realm of fuel cell-related electrocatalysis, CNT-based materials are assessed using CV, EIS, and polarization studies to determine their activity toward specific reactions such as oxygen reduction or hydrogen evolution [36]. Polarization curves generated in this context help identify the potential and current density associated with the reactions. These parameters are then compared across different CNT modifications to establish relative efficiencies. EIS can further unravel the interface impedance, while chronoamperometry tests under steady-state conditions evaluate long-term stability, which is vital in fuel cell environments [37]. Figure 2 illustrates a highly durable fuel cell electrocatalyst based on carbon nanotubes coated with double polymers [38].

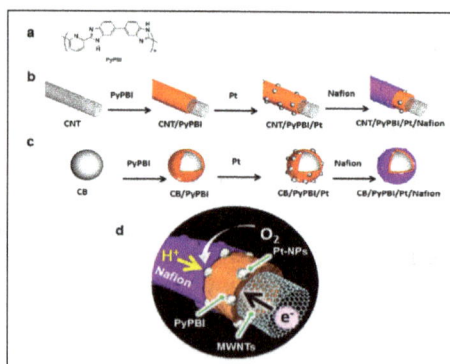

Figure 2. Schematic illustration of a highly durable fuel cell electrocatalyst based on double-polymer-coated carbon nanotubes (Reproduced with permission from Ref. [38],Copyright 2015, Springer Nature).

Electrocatalysts and Advanced Materials for Sustainable Energy Storage   Materials Research Forum LLC
Materials Research Foundations 182 (2025) 70-86   https://doi.org/10.21741/9781644903797-6

Applications of CNT-Based Catalysts in Energy Storage

Carbon nanotube (CNT) catalysts have attracted attention in energy storage and conversion technologies because of their exceptional electrical conductivity, extensive surface area, high mechanical strength, and distinctive chemical properties. These attributes make them highly effective as electrode materials, catalyst supports, and conductive additives in electrochemical systems. CNTs are crucial in enhancing charge transfer kinetics, improving reaction efficiency, and increasing device longevity, making them indispensable in **supercapacitors, lithium-ion/sodium-ion batteries, fuel cells, and hydrogen storage** [39][40]. This section delves into the key applications of CNT-based catalysts in **electrochemical energy storage** and **conversion systems**, highlighting their mechanisms, advantages, and challenges.

Supercapacitors: Supercapacitors, also referred to as electrochemical capacitors, are energy storage devices known for their high power density, fast charge-discharge cycles, and long cycle life. Unlike batteries, which store energy through chemical reactions, supercapacitors store energy by accumulating electrostatic charge at the interface between the electrode and electrolyte [41]. CNTs have emerged as promising materials for supercapacitor electrodes due to their:

   i.   High surface area for increased charge storage.

   ii.   Superior electrical conductivity for efficient charge transport.

   iii.   Mechanical stability for sustained performance.

Lithium-ion and Sodium-Ion Batteries: CNTs as electrode materials and conductive additives: Lithium-ion batteries (LIBs) and sodium-ion batteries (SIBs) are the backbone of modern energy storage for **portable electronics and electric vehicles**. CNTs improve battery performance by a) enhancing the electronic conductivity of electrodes, b) acting as structural buffers to prevent material degradation, and c) facilitating lithium/sodium-ion transport through interconnected nanostructures [42,43].

CNTs as electrode materials and conductive additives: Incorporating CNTs directly into electrode structures can substantially enhance the electrochemical performance of batteries. Their unique one-dimensional morphology facilitates efficient electron transport pathways and lays out a sturdy framework to accommodate the volumetric changes occurring during charge-discharge cycles. For instance, in lithium-ion batteries (LIBs), CNT-based electrodes have demonstrated improved capacity retention and rate capabilities. Another study highlights that electrodes incorporating multi-walled carbon nanotubes exhibit enhanced electrical conductivity and corrosion resistance, contributing to better battery performance [44,45].

Beyond serving as primary electrode materials, CNTs are extensively utilised as conductive additives to improve the electrical conductivity of electrode composites. Traditional conductive additives like carbon black are being supplemented or replaced by CNTs due to their superior conductive properties and the ability to form percolated networks at lower loadings. This substitution not only enhances electrical connectivity but also preserves the energy density of the electrode. Research indicates that including CNTs as conductive additives in LIB electrodes can significantly improve electrical and mechanical properties, which is particularly beneficial for flexible battery applications [46].

Hydrogen Evolution and Fuel Cells: Carbon nanotubes (CNTs) are essential components in fuel cells and hydrogen evolution reactions (HER) as they serve as conductive additives and electrode materials. In fuel cells, CNTs are effective catalyst supports owing to their high electrical conductivity, large surface area, and corrosion resistance, which improve electron transport and catalyst dispersion, particularly for platinum (Pt) or non-precious metal catalysts employed in the oxygen reduction reaction (ORR) at the cathode. Nitrogen-doped CNTs have been found to exhibit encouraging activity as independent electrocatalysts for ORR, offering a cost-effective solution compared to Pt-based materials [47-49]. In hydrogen evolution reactions, CNTs play a vital role in reducing the overpotential and enhancing the reaction kinetics when used with metal catalysts like MoS_2, Ni, or Co by acting as conductive scaffolds that allow fast electron transfer. Functionalised or doped CNTs have shown superior HER activity and stability under acidic and alkaline media. Their tunability and versatility in structure make CNTs a critical component in designing next-generation electrocatalytic systems for clean energy technologies [50-53].

Role of CNT-supported catalysts in fuel cell electrodes: Several benefits are witnessed in fuel cells with the support of CNTs, such as enhanced electrochemical corrosion resistance, higher catalytic activity, and the capability of working at elevated current densities as opposed to conventional catalyst supports. Moreover, CNTs help enhance mass transmission in both electrodes of the fuel cell system [54,55]. Nitrogen-doped carbon nanotubes (N-CNTs) have shown inherent catalytic activity for the oxygen reduction reaction (ORR), offering a cost-effective alternative to traditional platinum-based catalysts. Their hierarchical porous structures enhance mass transport and ion diffusion, leading to higher power densities in fuel cells. Consequently, CNT-supported catalysts strengthen next-generation fuel cells' performance, durability, and economic feasibility for clean energy applications [56,57].

CNT-based catalysts in hydrogen storage and electrolysis: Carbon nanotubes (CNTs) have demonstrated their potential in fabricating efficient heterogeneous catalysts for hydrogen production. Additionally, CNTs are a convenient adsorbent material, which could be the foundation for effective hydrogen storage systems [58]. Since their discovery, one-dimensional nanostructured materials have demonstrated promising properties as catalyst supports in fuel-cell applications. This is primarily due to their unique morphology and advantageous characteristics, which include a nanoscale size, a high accessible surface area, good electronic conductivity, and excellent stability. For instance, it has been reported that CNTs outperform carbon blacks as catalyst supports in proton exchange membrane fuel cells [59-62].

CNT-based catalysts possess a promising ability to advance the efficiency and longevity of H_2 production using electrolysis. Hydrogen evolution (HER) and oxygen evolution (OER), the essential reactions involved, need active and robust catalysts. Carbon nanotubes (CNTs) have high electrical conductivity, a huge surface area, and excellent mechanical strength. This ensures they are apt candidates for distributing catalytic phases such as transition metals (such as Ni, Co, Mo) and metal phosphides or sulphides. These composites increase the active surface area, facilitate quick electron transfer, and enhance stability under harsh conditions [63,64].

Challenges and Future Prospects

The major challenge in obtaining high-purity carbon nanotubes (CNTs) involves scalable and inexpensive synthesis. Such conventional techniques as chemical vapour deposition, arc discharge, and laser ablation involve costly devices and energy that utilise metal catalysts, introducing contamination and impacting performance. Another major technical issue is upholding uniformity in dimensions and morphology across different batches[65]. Functionalisation and doping are

Electrocatalysts and Advanced Materials for Sustainable Energy Storage   Materials Research Forum LLC
Materials Research Foundations 182 (2025) 70-86   https://doi.org/10.21741/9781644903797-6

essential for enhancing the catalytic properties of CNTs, but can inadvertently introduce defects that compromise electrical conductivity and structural integrity. Controlling the degree of functionalisation without sacrificing key properties is an ongoing materials science challenge. Additionally, the interfacial stability of CNT-based electrodes during repeated electrochemical cycling in harsh environments (acidic, alkaline, or high-voltage conditions) poses significant concerns for practical battery or supercapacitor applications [66].

From an electrochemical perspective, reproducibility in performance and the mechanistic understanding of CNT-assisted charge transfer processes require further study. In complex hybrid systems, it is often difficult to decouple the contribution of CNTs from that of co-catalysts or active materials [67,68]. Despite these issues, the future of CNT-based catalysts is very promising. New low-cost synthesis methods, including biomass-based carbon nanotubes and plasma-enhanced routes, hold promise for more environmentally friendly and scalable manufacturing. Improved computational modelling and machine learning will also be anticipated to speed up the development of optimised CNT architectures and heterostructures suitable for particular electrochemical reactions [69-71].

In addition, hybrid composites like CNT-graphene, CNT-metal oxide, and CNT-conducting polymers are increasingly being developed based on their synergistic properties. Heteroatom doping of CNTs (N, B, S, or P) also continues to explore new avenues for high-performance catalysis. Finally, 3D printing and flexible electronics bring new application fields for CNT-based materials into wearable energy storage devices and portable power systems. Although technical and economic challenges are significant, consistent innovation in synthesis, characterisation, and application design will position CNT-based catalysts at the forefront of energy storage systems in the near future [72-74].

Conclusion

Carbon nanotubes (CNTs) are highly versatile and effective materials for creating advanced catalysts used in energy storage and conversion systems. Among the available synthesis methods, Chemical Vapour Deposition (CVD) is recognised as a highly cost-effective and scalable technique, offering precise control over CNT structure and yield. Modifications like surface functionalisation, doping with heteroatoms or metals, and the formation of composite materials significantly improve the catalytic, electronic, and mechanical properties of carbon nanotubes. The engineered enhancements enable CNTs to perform exceptionally in various roles, from catalyst supports to active electrode materials. Electrochemical techniques, including cyclic voltammetry, electrochemical impedance spectroscopy, and Tafel analysis, provide critical insights into the performance and stability of CNT-based catalysts. These evaluations help understand the charge transfer mechanisms and guide the optimisation of CNT materials for specific applications. CNT-based catalysts are already pivotal in energy technologies, including supercapacitors, lithium-ion and sodium-ion batteries, fuel cells, and hydrogen evolution systems. Their high surface area, conductivity, and structural resilience make them ideal for improving efficiency and durability in these systems. Nevertheless, challenges remain regarding scalable synthesis, purity control, and maintaining consistent electrochemical performance. Future progress will likely depend on innovations in fabrication techniques, deeper mechanistic understanding, and the development of hybrid materials. As research advances, CNT-based catalysts are expected to be central to the evolution of high-performance, sustainable energy storage solutions.

References

[1] M. Li, Z. Li, Q. Lin, J. Cao, F. Liu, S. Kawi, Synthesis strategies of carbon nanotube supported and confined catalysts for thermal catalysis, Chemical Engineering Journal 431 (2022) 133970. https://doi.org/10.1016/j.cej.2021.133970

[2] C. Sathiskumar, S. Karthikeyan, Recycling of waste tires and its energy storage application of by-products –a review, Sustainable Materials and Technologies 22 (2019) e00125. https://doi.org/10.1016/j.susmat.2019.e00125

[3] P. Misra, D. Casimir, R. Garcia-Sanchez, Raman Spectroscopy and Molecular Dynamics Simulation Studies of Carbon Nanotubes, in: Environmental Science and Engineering, Elsevier, 2014: pp. 507–510. https://doi.org/10.1007/978-3-319-03002-9_127

[4] R. Sharma, A.K. Sharma, V. Sharma, Synthesis of carbon nanotubes by arc-discharge and chemical vapor deposition method with analysis of its morphology, dispersion and functionalization characteristics, Cogent Engineering 2 (2015) 1094017. https://doi.org/10.1080/23311916.2015.1094017

[5] M. Kim, S. Osone, T. Kim, H. Higashi, T. Seto, Synthesis of nanoparticles by laser ablation: A review, KONA Powder and Particle Journal 2017 (2017) 80–90. https://doi.org/10.14356/kona.2017009

[6] Y. Sani, R.S. Azis, I. Ismail, Y. Yaakob, J. Mohammed, Enhanced electromagnetic microwave absorbing performance of carbon nanostructures for RAMs: A review, Applied Surface Science Advances 18 (2023) 100455. https://doi.org/10.1016/j.apsadv.2023.100455

[7] Y. Sani, R.S. Azis, I. Ismail, Y. Yaakob, J. Mohammed, Enhanced electromagnetic microwave absorbing performance of carbon nanostructures for RAMs: A review, Applied Surface Science Advances 18 (2023) 100455. https://doi.org/10.1016/j.apsadv.2023.100455

[8] X. Zeng, X. Cheng, R. Yu, G.D. Stucky, Electromagnetic microwave absorption theory and recent achievements in microwave absorbers, Carbon 168 (2020) 606–623. https://doi.org/10.1016/j.carbon.2020.07.028

[9] D. Khalafallah, R. Sarkar, M. Demir, K.A. Khalil, Z. Hong, A.A. Farghaly, Heteroatoms-doped carbon nanotubes for energy applications, in: Handbook of Carbon Nanotubes, Springer, 2022: pp. 485–523. https://doi.org/10.1007/978-3-030-91346-5_68

[10] K. Sheoran, H. Kaur, S.S. Siwal, A.K. Saini, D.V.N. Vo, V.K. Thakur, Recent advances of carbon-based nanomaterials (CBNMs) for wastewater treatment: Synthesis and application, Chemosphere 299 (2022) 134364. https://doi.org/10.1016/j.chemosphere.2022.134364

[11] N. Puri, A. Gupta, A review on design and mechanism approaches of metal-organic framework composites for supercapacitance performance, Coordination Chemistry Reviews 540 (2025) 216772. https://doi.org/10.1016/j.ccr.2025.216772

[12] A. Tundwal, H. Kumar, B.J. Binoj, R. Sharma, G. Kumar, R. Kumari, A. Dhayal, A. Yadav, D. Singh, P. Kumar, Developments in conducting polymer-, metal oxide-, and carbon nanotube-based composite electrode materials for supercapacitors: a review, RSC Advances 14 (2024) 9406–9439. https://doi.org/10.1039/d3ra08312h

[13]   L.R. Bard, Allen J., Faulkner, Fundamentals and Applications Plasmonics : Fundamentals and Applications, John Wiley & Sons, 2001.

[14]   S. Bashir, P. Hanumandla, H.Y. Huang, J.L. Liu, Nanostructured materials for advanced energy conversion and storage devices: Safety implications at end-of-life disposal, Nanostructured Materials for Next-Generation Energy Storage and Conversion: Fuel Cells 4 (2018) 517–542. https://doi.org/10.1007/978-3-662-56364-9_18

[15]   Y. Yan, J. Miao, Z. Yang, F.X. Xiao, H. Bin Yang, B. Liu, Y. Yang, Carbon nanotube catalysts: Recent advances in synthesis, characterization and applications, Chemical Society Reviews 44 (2015) 3295–3346. https://doi.org/10.1039/c4cs00492b

[16]   Z. Qi, Electrochemical methods for catalyst activity evaluation, in: PEM Fuel Cell Electrocatalysts and Catalyst Layers: Fundamentals and Applications, Springer, 2008: pp. 547–607. https://doi.org/10.1007/978-1-84800-936-3_11

[17]   D. Diamond, Analytical electrochemistry, TrAC Trends in Analytical Chemistry 15 (1996) X–XI. https://doi.org/10.1016/s0165-9936(96)90116-8

[18]   E.M. Espinoza, J.A. Clark, J. Soliman, J.B. Derr, M. Morales, V.I. Vullev, Practical Aspects of Cyclic Voltammetry: How to Estimate Reduction Potentials When Irreversibility Prevails, Journal of The Electrochemical Society 166 (2019) H3175–H3187. https://doi.org/10.1149/2.0241905jes

[19]   H. Beitollahi, F. Movahedifar, S. Tajik, S. Jahani, A Review on the Effects of Introducing CNTs in the Modification Process of Electrochemical Sensors, Electroanalysis 31 (2019) 1195–1203. https://doi.org/10.1002/elan.201800370

[20]   R. Kulkarni, L.P. Lingamdinne, J.R. Koduru, R.R. Karri, S.K. Kailasa, N.M. Mubarak, Y.Y. Chang, M.H. Dehghani, Exploring the recent cutting-edge applications of CNTs in energy and environmental remediation: Mechanistic insights and remarkable performance advancements, Journal of Environmental Chemical Engineering 12 (2024) 113251. https://doi.org/10.1016/j.jece.2024.113251

[21]   K. Ariyoshi, A. Mineshige, M. Takeno, T. Fukutsuka, T. Abe, S. Uchida, Z. Siroma, Electrochemical Impedance Spectroscopy Part 2: Applications†, Electrochemistry 90 (2022) 102008. https://doi.org/10.5796/electrochemistry.22-66080

[22]   전종한, Study on FC-CVD-Synthesized CNT Assemblies for Energy Storage and Conversion, (2023).

[23]   M. Tertiş, A. Florea, B. Feier, I.O. Marian, L. Silaghi-Dumitrescu, A. Cristea, R. Sandulescu, C. Cristea, Electrochemical impedance studies on single and multi-walled carbon nanotubes - Polymer nanocomposites for biosensors development, Journal of Nanoscience and Nanotechnology 15 (2015) 3385–3393. https://doi.org/10.1166/jnn.2015.10208

[24]   N.E. Tolouei, S. Ghamari, M. Shavezipur, Development of circuit models for electrochemical impedance spectroscopy (EIS) responses of interdigitated MEMS biochemical sensors, Journal of Electroanalytical Chemistry 878 (2020) 114598. https://doi.org/10.1016/j.jelechem.2020.114598

[25]   Y.S. Choudhary, L. Jothi, G. Nageswaran, Electrochemical Characterization, in: Spectroscopic Methods for Nanomaterials Characterization, Elsevier, 2017: pp. 19–54. https://doi.org/10.1016/B978-0-323-46140-5.00002-9

[26]   H.J. Engell, Stability and breakdown phenomena of passivating films, Electrochimica Acta 22 (1977) 987–993. https://doi.org/10.1016/0013-4686(77)85010-X

[27]   T. Shinagawa, A.T. Garcia-Esparza, K. Takanabe, Insight on Tafel slopes from a microkinetic analysis of aqueous electrocatalysis for energy conversion, Scientific Reports 5 (2015) 13801. https://doi.org/10.1038/srep13801

[28]   N.B. Mansor, Development of Catalysts and Catalyst Supports for Polymer Electrolyte Fuel Cells, (2014) 164.

[29]   L.R. Bard, Allen J., Faulkner, Fundamentals and Applications Plasmonics : Fundamentals and Applications, John Wiley & Sons, 2001.

[30]   J. Ni, Y. Li, Carbon Nanomaterials in Different Dimensions for Electrochemical Energy Storage, Advanced Energy Materials 6 (2016) 1600278. https://doi.org/10.1002/aenm.201600278

[31]   K. Jiang, R.A. Gerhardt, Fabrication and Supercapacitor Applications of Multiwall Carbon Nanotube Thin Films, C 7 (2021) 70. https://doi.org/10.3390/c7040070

[32]   P. Simon, Y. Gogotsi, Materials for electrochemical capacitors, Materials for Sustainable Energy: A Collection of Peer-Reviewed Research and Review Articles from Nature Publishing Group 7 (2010) 138–147. https://doi.org/10.1142/9789814317665_0021

[33]   U. Nisar, R.A. Shakoor, R. Essehli, R. Amin, B. Orayech, Z. Ahmad, P.R. Kumar, R. Kahraman, S. Al-Qaradawi, A. Soliman, Sodium intercalation/de-intercalation mechanism in Na4MnV(PO4)3 cathode materials, Electrochimica Acta 292 (2018) 98–106. https://doi.org/10.1016/j.electacta.2018.09.111

[34]   S.K. Rath, S. Dubey, G.S. Kumar, S. Kumar, A.K. Patra, J. Bahadur, A.K. Singh, G. Harikrishnan, T.U. Patro, Multi-walled CNT-induced phase behaviour of poly (vinylidene fluoride) and its electro-mechanical properties, Journal of Materials Science 49 (2014) 103–113.

[35]   G. Wu, P. Tan, D. Wang, Z. Li, L. Peng, Y. Hu, C. Wang, W. Zhu, S. Chen, W. Chen, High-performance Supercapacitors Based on Electrochemical-induced Vertical-aligned Carbon Nanotubes and Polyaniline Nanocomposite Electrodes, Scientific Reports 7 (2017) 43676. https://doi.org/10.1038/srep43676

[36]   S. Trasatti, O.A. Petrii, Real surface area measurements in electrochemistry, Journal of Electroanalytical Chemistry 327 (1992) 353–376. https://doi.org/10.1016/0022-0728(92)80162-W

[37]   H. Liu, C. Song, L. Zhang, J. Zhang, H. Wang, D.P. Wilkinson, A review of anode catalysis in the direct methanol fuel cell, Journal of Power Sources 155 (2006) 95–110. https://doi.org/10.1016/j.jpowsour.2006.01.030

[38]   M.R. Berber, I.H. Hafez, T. Fujigaya, N. Nakashima, A highly durable fuel cell electrocatalyst based on double-polymer-coated carbon nanotubes, Scientific Reports 5 (2015) 16711. https://doi.org/10.1038/srep16711

[39]   T.W. Chen, S.M. Chen, G. Anushya, R. Kannan, P. Veerakumar, A.G. Al-Sehemi, V. Mariyappan, S. Alargarsamy, M.M. Alam, T.C. Mahesh, R. Ramachandran, P. Kalimuthu, Electrochemical energy storage applications of functionalized carbon-based nanomaterials: An overview, International Journal of Electrochemical Science 19 (2024) 100548. https://doi.org/10.1016/j.ijoes.2024.100548

[40]   A.T. Lawal, Recent application of carbon nanotube in energy storage devices, Fujnas (2024) 100470.

[41]   G.B. Pour, H. Ashourifar, L.F. Aval, S. Solaymani, CNTs-Supercapacitors: A Review of Electrode Nanocomposites Based on CNTs, Graphene, Metals, and Polymers, Symmetry 15 (2023) 1179. https://doi.org/10.3390/sym15061179

[42]   G. Wang, H. Li, Q. Zhang, Z. Yu, M. Qu, The study of carbon nanotubes as conductive additives of cathode in lithium ion batteries, Journal of Solid State Electrochemistry 15 (2011) 759–764. https://doi.org/10.1007/s10008-010-1143-4

[43]   F. Tao, Y. Liu, X. Ren, A. Jiang, H. Wei, X. Zhai, F. Wang, H.R. Stock, S. Wen, F. Ren, Carbon nanotube-based nanomaterials for high-performance sodium-ion batteries: Recent advances and perspectives, Journal of Alloys and Compounds 873 (2021) 159742. https://doi.org/10.1016/j.jallcom.2021.159742

[44]   F. Tao, Y. Liu, X. Ren, A. Jiang, H. Wei, X. Zhai, F. Wang, H.R. Stock, S. Wen, F. Ren, Carbon nanotube-based nanomaterials for high-performance sodium-ion batteries: Recent advances and perspectives, Journal of Alloys and Compounds 873 (2021) 159742. https://doi.org/10.1016/j.jallcom.2021.159742

[45]   N. V. Glebova, A.A. Nechitailov, A. Krasnova, Electrode material containing carbon nanotubes and its kinetic characteristics of oxygen electroreduction, Reaction Kinetics, Mechanisms and Catalysis 131 (2020) 599–612. https://doi.org/10.1007/s11144-020-01866-w

[46]   S. Jessl, D. Beesley, S. Engelke, C.J. Valentine, J.C. Stallard, N. Fleck, S. Ahmad, M.T. Cole, M. De Volder, Carbon nanotube conductive additives for improved electrical and mechanical properties of flexible battery electrodes, Materials Science and Engineering: A 735 (2018) 269–274. https://doi.org/10.1016/j.msea.2018.08.033

[47]   X. Zeng, S. Shahgaldi, S.K. Mitra, X. Li, Impact of carbon supports on the Pt-based catalyst activity and fuel cell performance under varied operational conditions, Energy Conversion and Management 326 (2025) 119496. https://doi.org/10.1016/j.enconman.2025.119496

[48]   J.D. Sinniah, W.Y. Wong, K.S. Loh, R.M. Yunus, S.N. Timmiati, Perspectives on carbon-alternative materials as Pt catalyst supports for a durable oxygen reduction reaction in proton exchange membrane fuel cells, Journal of Power Sources 534 (2022) 231422. https://doi.org/10.1016/j.jpowsour.2022.231422

[49]   P. Agrawal, S. Ebrahim, D. Ponnamma, Advancements in nanocarbon-based catalysts for enhanced fuel cell performance: a comprehensive review, International Journal of Energy and Water Resources (2024) 1–23. https://doi.org/10.1007/s42108-024-00324-w

[50]   T. Rasheed, S. Munir, A. BaQais, M. Shahid, M.A. Amin, M.F. Warsi, S. Yousaf, Multi-walled carbon nanotubes with embedded nickel sulphide as an effective electrocatalyst for Bi-functional water splitting, International Journal of Hydrogen Energy 67 (2024) 373–380. https://doi.org/10.1016/j.ijhydene.2024.04.131

[51]   S. Majumdar, S. Chaitoglou, J. Serafin, G. Farid, R. Ospina, Y. Ma, R. Amade Rovira, E. Bertran-Serra, Enhancing hydrogen evolution: Carbon nanotubes as a scaffold for Mo2C deposition via magnetron sputtering and chemical vapor deposition, International Journal of Hydrogen Energy 89 (2024) 977–989. https://doi.org/10.1016/j.ijhydene.2024.09.425

[52]   S. Li, E. Li, X. An, X. Hao, Z. Jiang, G. Guan, Transition metal-based catalysts for electrochemical water splitting at high current density: Current status and perspectives, Nanoscale 13 (2021) 12788–12817. https://doi.org/10.1039/d1nr02592a

[53]   P. Tripathi, A.K. Verma, A.S.K. Sinha, S. Singh, Graphene-Transition Metal Electrocatalysts for Sustainable Water Electrolysis, ChemistrySelect 9 (2024) e202403155. https://doi.org/10.1002/slct.202403155

[54]   P. Agrawal, S. Ebrahim, D. Ponnamma, Advancements in nanocarbon-based catalysts for enhanced fuel cell performance: a comprehensive review, International Journal of Energy and Water Resources (2024) 1–23. https://doi.org/10.1007/s42108-024-00324-w

[55]   R. Sharma, M. Almáši, R.C. Punia, R. Chaudhary, S.P. Nehra, M.S. Dhaka, A. Sharma, Solar-driven polymer electrolyte membrane fuel cell for photovoltaic hydrogen production, International Journal of Hydrogen Energy 48 (2023) 37999–38014. https://doi.org/10.1016/j.ijhydene.2022.12.175

[56]   X. Wang, Z. Kong, J. Ye, C. Shao, B. Li, Hollow nitrogen-doped carbon nanospheres as cathode catalysts to enhance oxygen reduction reaction in microbial fuel cells treating wastewater, Environmental Research 201 (2021) 111603. https://doi.org/10.1016/j.envres.2021.111603

[57]   Y. Nie, X. Xie, S. Chen, W. Ding, X. Qi, Y. Wang, J. Wang, W. Li, Z. Wei, M. Shao, Towards Effective Utilization of Nitrogen-Containing Active Sites: Nitrogen-doped Carbon Layers Wrapped CNTs Electrocatalysts for Superior Oxygen Reduction, Electrochimica Acta 187 (2016) 153–160. https://doi.org/10.1016/j.electacta.2015.11.011

[58]   R. Oriňáková, A. Oriňák, Recent applications of carbon nanotubes in hydrogen production and storage, Fuel 90 (2011) 3123–3140. https://doi.org/10.1016/j.fuel.2011.06.051

[59]   R. Oriňáková, A. Oriňák, Recent applications of carbon nanotubes in hydrogen production and storage, Fuel 90 (2011) 3123–3140. https://doi.org/10.1016/j.fuel.2011.06.051

[60]   J. Deng, X. Ding, W. Zhang, Y. Peng, J. Wang, X. Long, P. Li, A.S.C. Chan, Carbon nanotube–polyaniline hybrid materials, European Polymer Journal 38 (2002) 2497–2501.

[61]   Y. Liang, H. Zhang, B. Yi, Z. Zhang, Z. Tan, Preparation and characterization of multi-walled carbon nanotubes supported PtRu catalysts for proton exchange membrane fuel cells, Carbon 43 (2005) 3144–3152. https://doi.org/10.1016/j.carbon.2005.06.017

[62] C. Wang, M. Waje, X. Wang, J.M. Tang, R.C. Haddon, Y. Yan, Proton Exchange Membrane Fuel Cells with Carbon Nanotube Based Electrodes, Nano Letters 4 (2004) 345–348. https://doi.org/10.1021/nl034952p

[63] N.A. Kamaruzaman, W.M. Khairul, N. Md Saleh, F. Yusoff, Advancements in carbon-based transition metal compounds for enhanced hydrogen production via electrochemical water splitting, International Journal of Electrochemical Science 19 (2024) 100740. https://doi.org/10.1016/j.ijoes.2024.100740

[64] G.T.M. Kadja, M. Mualliful Ilmi, S. Mardiana, M. Khalil, F. Sagita, N.T.U. Culsum, A.T.N. Fajar, Recent advances of carbon nanotubes as electrocatalyst for in-situ hydrogen production and CO2 conversion to fuels, Results in Chemistry 6 (2023) 101037. https://doi.org/10.1016/j.rechem.2023.101037

[65] A. Kaushal, R. Alexander, J. Prakash, K. Dasgupta, Engineering challenges and innovations in controlled synthesis of CNT fiber and fabrics in floating catalyst chemical vapor deposition (FC-CVD) process, Diamond and Related Materials 148 (2024) 111474. https://doi.org/10.1016/j.diamond.2024.111474

[66] H. Li, G. Wang, Y. Wu, N. Jiang, K. Niu, Functionalization of Carbon Nanotubes in Polystyrene and Properties of Their Composites: A Review, Polymers 16 (2024) 770. https://doi.org/10.3390/polym16060770

[67] J. Dugeč, I. Škugor Rončević, N. Vladislavić, J. Radić, M. Buljac, M. Buzuk, The Interpretation of Carbon Nanotubes' Electrochemistry: Electrocatalysis and Mass Transport Regime in the Apparent Promotion of Electron Transfer, Biosensors 15 (2025) 89. https://doi.org/10.3390/bios15020089

[68] R. Rao, C.L. Pint, A.E. Islam, R.S. Weatherup, S. Hofmann, E.R. Meshot, F. Wu, C. Zhou, N. Dee, P.B. Amama, J. Carpena-Nuñez, W. Shi, D.L. Plata, E.S. Penev, B.I. Yakobson, P.B. Balbuena, C. Bichara, D.N. Futaba, S. Noda, H. Shin, K.S. Kim, B. Simard, F. Mirri, M. Pasquali, F. Fornasiero, E.I. Kauppinen, M. Arnold, B.A. Cola, P. Nikolaev, S. Arepalli, H.M. Cheng, D.N. Zakharov, E.A. Stach, J. Zhang, F. Wei, M. Terrones, D.B. Geohegan, B. Maruyama, S. Maruyama, Y. Li, W.W. Adams, A.J. Hart, Carbon Nanotubes and Related Nanomaterials: Critical Advances and Challenges for Synthesis toward Mainstream Commercial Applications, ACS Nano 12 (2018) 11756–11784. https://doi.org/10.1021/acsnano.8b06511

[69] M. El-Azazy, A.I. Osman, M. Nasr, Y. Ibrahim, N. Al-Hashimi, K. Al-Saad, M.A. Al-Ghouti, M.F. Shibl, A.H. Al-Muhtaseb, D.W. Rooney, A.S. El-Shafie, The interface of machine learning and carbon quantum dots: From coordinated innovative synthesis to practical application in water control and electrochemistry, Coordination Chemistry Reviews 517 (2024). https://doi.org/10.1016/j.ccr.2024.215976

[70] G. Chen, D.M. Tang, Machine Learning as a "Catalyst" for Advancements in Carbon Nanotube Research, Nanomaterials 14 (2024) 1688. https://doi.org/10.3390/nano14211688

[71] A.K. Madikere Raghunatha Reddy, A. Darwiche, M.V. Reddy, K. Zaghib, Review on Advancements in Carbon Nanotubes: Synthesis, Purification, and Multifaceted Applications †, Batteries 11 (2025). https://doi.org/10.3390/batteries11020071

[72]   M. Tavakkoli, E. Flahaut, P. Peljo, J. Sainio, F. Davodi, E. V. Lobiak, K. Mustonen, E.I. Kauppinen, Mesoporous Single-Atom-Doped Graphene-Carbon Nanotube Hybrid: Synthesis and Tunable Electrocatalytic Activity for Oxygen Evolution and Reduction Reactions, ACS Catalysis 10 (2020) 4647–4658. https://doi.org/10.1021/acscatal.0c00352

[73]   F. Davodi, M. Tavakkoli, J. Lahtinen, T. Kallio, Straightforward synthesis of nitrogen-doped carbon nanotubes as highly active bifunctional electrocatalysts for full water splitting, Journal of Catalysis 353 (2017) 19–27. https://doi.org/10.1016/j.jcat.2017.07.001

[74]   B. Liu, J. Sun, J. Zhao, X. Yun, Hybrid graphene and carbon nanotube–reinforced composites: polymer, metal, and ceramic matrices, Advanced Composites and Hybrid Materials 8 (2025) 1. https://doi.org/10.1007/s42114-024-01074-3

Electrocatalysts and Advanced Materials for Sustainable Energy Storage   Materials Research Forum LLC
Materials Research Foundations 182 (2025) 87-99   https://doi.org/10.21741/9781644903797-7

Chapter 7

Biomass-Derived Carbon Materials: Harnessing Electrodes for Sustainable Energy Storage Devices

Priyadharsini Natarajan[1,a*], Subashini Chinnasamy[1,b], Keerthana Prabhakaran[1,c]

[1]Department of Physics, PSGR Krishnammal College for Women, Coimbatore – 641004, Tamil Nadu, India

[a]priyadharsinin@psgrkcw.ac.in, [b]csubashini1995@gmail.com, [c]chellamakeerthi@gmail.com

Abstract

The shift toward sustainable energy systems necessitates the expansion of efficient and eco-friendly carbon materials for energy storage and conversion. Biomass-derived carbon materials have attracted considerable interest owing to their sustainability, adjustable physicochemical properties, and economic feasibility. Derived from agricultural waste, lignocellulosic biomass, and other organic sources, these materials exhibit unique structural and chemical characteristics, such as a high surface area, heteroatom doping, and tunable porosity, that enhance their electrochemical performance. This chapter explores the role of biomass-derived carbon materials in energy storage technologies, with a particular focus on their applications in batteries and supercapacitors. A range of synthesis methods, including carbonization, activation, and functionalization, is discussed to illustrate how electrochemical properties can be optimized for improved performance. Additionally, the electrochemical efficiency and durability of these carbon materials are evaluated in comparison to conventional high-cost alternatives. Biomass-derived carbon materials contribute to a reduced carbon footprint, promote resource circularity, and facilitate the valorization of waste. Their scalability and commercialization potential are examined by considering factors such as raw material availability, processing challenges, and long-term stability. Biomass-derived carbon materials present a promising pathway toward more sustainable and efficient energy storage solutions by integrating green chemistry principles with advanced materials engineering.

Keywords

Biomass, Energy Storage, Supercapacitors, Porous Carbon, Batteries

Contents

Introduction

The increasing global reliance on fossil fuels and the resulting environmental consequences have significantly heightened the importance of advanced energy storage systems. Pollution associated with conventional energy sources not only accelerates climate change but also poses serious risks to public health. In response to these challenges, extensive research efforts are underway to advance renewable energy technologies and facilitate the transition toward a more sustainable and environmentally responsible energy landscape. However, renewable energy systems face inherent limitations, including intermittency, grid integration challenges, high capital costs, land-use constraints, and finite material availability. To address these obstacles, large-scale energy storage technologies such as hydrogen storage, pumped hydroelectric systems, utility-scale batteries, and compressed air energy storage are being explored as critical enablers of clean energy deployment. Among the various materials considered for energy storage, carbon-based materials have attracted considerable attention due to their favorable properties, including high surface area, tunable porosity, structural robustness, excellent electrical conductivity, and abundant availability. These attributes make them suitable for technical applications and economic scalability in sustainable energy systems. Biomass-derived carbon materials offer additional advantages by leveraging renewable feedstocks and aligning with green chemistry principles and circular economy goals, thereby promoting waste valorization and environmental sustainability [1][2]. Batteries and supercapacitors have emerged as promising candidates, with hard carbon (HC) playing a central role as a sustainable and high-performance electrode material. This review focuses on recent advances, challenges, and future perspectives of hard carbon in batteries and supercapacitors, particularly those derived from biomass.

Recent progress in energy storage research has underscored the potential of biomass-derived carbon materials as sustainable and cost-effective alternatives for commercial and next-generation battery systems. These materials offer numerous advantages, including environmental compatibility, abundant availability, and tunable physicochemical properties well-suited for electrochemical applications. Porous carbon structures derived from biomass play a crucial role in enhancing battery performance. Their high surface area and interconnected pore networks facilitate efficient ion diffusion and reversible ion intercalation/deintercalation, thereby improving charge–discharge characteristics and cycling stability [3]. Furthermore, the inherent thermal stability of these carbon materials supports long-term cycling performance by minimizing capacity degradation over extended use.

Biomass-derived carbon electrodes are more attractive due to their unique characteristics of sustainability, intrinsic conductivity, and tunable porosity, making them suitable for energy storage device applications. Its advantages are particularly present in the creation of a stable solid electrolyte interphase (SEI), which plays a significant role in various parameters, including maintaining electrode integrity, reducing electrolyte decomposition, and ensuring stable ion diffusion during prolonged cycling. These carbons also exhibit robust mechanical stability and structural resilience, while their surface functionalities can introduce additional electrochemically active sites, contributing to enhanced capacitive behavior. When hybridized with redox-active materials such as metal oxides or sulfides, they enable synergistic performance improvements [2]. Supercapacitors (SCs), emerged as a frontrunner among energy storage devices due to their superior power delivery, rapid charge–discharge ratio, and better cycling stability. In supercapacitors, carbon-based electrodes dominate the electric double-layer capacitance (EDLC) model of energy storage mechanism, which is a non-Faradaic process involving electrostatic ion accumulation at the electrode–electrolyte interface, ensuring high stability and efficiency over extended operation [4].

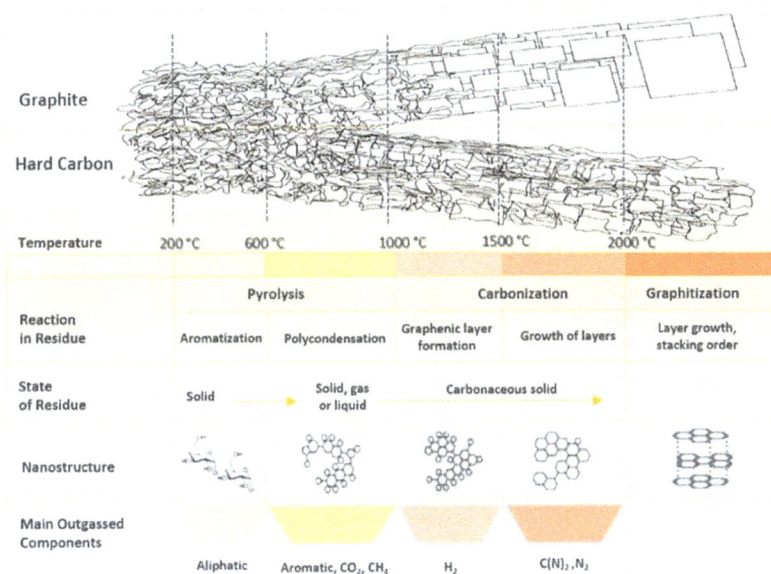

Figure 1. Hard Carbon Formation Pathway with Temperature: Circles Indicate Functional Groups and Side Chains in Polymeric Structures (Reproduced with permission from Ref. [6], Copyright 2019, Elsevier).

Hard carbon (HC), a non-graphitizable amorphous carbon typically derived from biomass, has garnered significant attention as a versatile electrode material for lithium-ion and sodium-ion energy storage systems. Its inherently disordered structure, enlarged interlayer spacing, and abundant surface functionalities allow for efficient and reversible storage of both lithium and sodium ions. These characteristics make HC especially attractive for hybrid capacitor systems and sodium-ion capacitors (NICs). In lithium-ion hybrid capacitors (LIHCs), the HC serves as a low-potential, insertion-type anode paired with high-surface-area, capacitive cathodes, achieving a balanced combination of high energy and power densities. Similarly, in NICs, HC competes with or complements other promising materials such as $Na_2Ti_3O_7$ and V_2O_5, offering stable cycling, fast ion kinetics, and structural integrity [5].

Figure 1 illustrates the temperature-dependent structural evolution of hard carbon and graphite during thermal processing, including pyrolysis, carbonization, and graphitization, highlighting changes in microstructure, volatile release, and residual composition [6]. The use of biomass waste as a precursor for HC production presents multiple advantages, including addressing issues of agricultural and organic waste management, reducing raw material costs, and contributing to the sustainability of electrode manufacturing. A variety of biomass sources, including lotus stems, soybean roots, rice husks, puffed rice, silk fibers, and popcorn, have been successfully carbonized into high-performance hard carbon materials [7]. These biomass-derived HCs often exhibit a synergistic combination of electric double-layer capacitance (EDLC) and pseudocapacitance, enabling enhanced ion transport, rapid charge–discharge capability, and prolonged cycling life. Such features make them highly suitable for next-generation, eco-friendly, and cost-effective energy storage technologies [8].

Hard Carbon in Battery Applications

Biomass-derived hard carbons are highly versatile for various battery technologies, including lithium-ion, sodium-ion, potassium-ion, and zinc-air batteries. Hard carbons synthesized from biomass sources exhibit broad applicability across multiple battery systems. Their multifunctional role, like spanning electron conduction, mechanical reinforcement, and interfacial stabilization, positions them as promising materials for advancing high-performance, eco-friendly energy storage systems [9,10].

Biomass-Derived Carbon Materials:

Lithium-Ion Batteries (LIBs): Since the commercial introduction of lithium-ion batteries (LIBs) in 1991, they have become a pivotal technology for energy storage. Lithium-ion batteries (LIBs) are widely preferred due to their high energy density, which makes them particularly suitable for use in electric vehicles (EVs). Traditionally, natural graphite has been employed as the anode material due to its layered structure, which allows for efficient lithium-ion intercalation [9]. However, as reported by Wu et al., natural graphite is limited by structural defects, microporosity, and partial sp^3 hybridization, despite offering a reasonable theoretical capacity of 335 mAh g^{-1} and an initial coulombic efficiency of approximately 88% [11].

Biomass-derived carbon materials (BCMs) served as highly effective precursors for the development of graphite-based anodes, owing to their sustainability, environmental compatibility, and economic advantages. The intrinsic porosity of BCMs enhances lithium-ion transport kinetics, facilitating stable cycling behavior. Furthermore, surface functionalization such as nitrogen or oxygen doping can improve electrical conductivity, increase active sites, and enhance capacity retention. Although transition metal oxides are employed as anode materials, they often encounter challenges such as considerable volume changes and structural instability during repeated charge–discharge cycles. Biomass-derived porous carbons, with their mechanical flexibility and tunable porosity, have demonstrated the ability to buffer these volume changes, thereby enhancing the overall structural integrity and electrochemical performance of composite electrodes. Nitrogen-doped porous carbon nanotubes derived from reed catkins flowers were studied by Zhang et al. This porous carbon electrode has a high surface area of 1751.1 m^2 g^{-1}. The material exhibited an impressive specific capacity of 2016.3 mAh g^{-1} at 0.02 A g^{-1} and maintained 512.4 mAh g^{-1} after 500 cycles, demonstrating excellent stability. Compared to other biomass-derived carbons reported in literature, such as those from coconut or peanut shells, the reed-based carbon electrode exhibited superior electrochemical performance, making it a strong candidate for high-performance lithium-ion battery anodes. Its structural and chemical features significantly improve long-term cycling and rate capabilities [12].

Moreover, hard carbon demonstrates excellent thermal stability, with resistance to degradation at temperatures as high as 3000 °C, making it a potential candidate for use as a surface coating or dopant in lithium-ion battery anodes. Hard carbon offers several advantages over natural graphite, including higher reversible capacity, improved initial coulombic efficiency, simpler synthesis procedures, and reduced production cost. These results collectively underscore the suitability of biomass-derived carbon materials as sustainable and effective alternatives to traditional anode materials in lithium-ion batteries, offering solutions to both performance-related and environmental concerns [13].

Sodium-Ion Batteries (SIBs): The increasing dependence on portable electronic devices with rapid charging capabilities has intensified lithium consumption, raising concerns regarding its limited availability and escalating market prices. As a result, SIBs have become a popular area of research because sodium is widely available, low-cost, and more environmentally friendly. While sodium behaves similarly to lithium-ion batteries, its larger ion size (1.02 Å) necessitates the development of electrode materials that can accommodate this size difference.

HC has emerged as a suitable anode for SIBs. It has high porosity, expanded interlayer spacing, and structural disorder, which collectively enhance sodium-ion storage and transport kinetics. Typically synthesized via pyrolysis of biomass and other carbon-rich precursors such as agricultural waste, phenolic resins, and wood by-products, HC maintains the morphological features of its source material, offering advantageous closed-pore architectures conducive to sodium storage [10,14].

Uldana et al. investigated hard carbon (HC) derived from buckwheat, revealing a well-developed hierarchical mesoporous structure that facilitates improved electrolyte infiltration and enhances sodium-ion transport kinetics. A pre-oxidized sample, subsequently carbonized at 1400 °C, demonstrated a notable reversible capacity of 330.23 mAh g^{-1} and excellent coulombic efficiency ranging from 97% to 98% over 100 charge–discharge cycles. These results emphasize the importance of thermal pre-treatment in optimizing pore structure and surface chemistry to improve electrochemical performance [15]. Similarly, Yu et al. synthesized HC from old loofah through pyrolysis at 800 °C. The resulting material exhibited an irregular microstructure conducive to enhanced cycling stability. It delivered a high initial discharge capacity of 695 mAh g^{-1} at 25 mA g^{-1} and retained 171 mAh g^{-1} even after 1000 cycles at a high current density of 1000 mA g^{-1}. These studies collectively reinforce the potential of biomass-derived hard carbons as sustainable, cost-effective, and high-performance anode materials for next-generation sodium-ion batteries, in alignment with circular economy and environmental sustainability goals [16].Figure 2 represents the mechanism of hard carbon in SIBs [17].

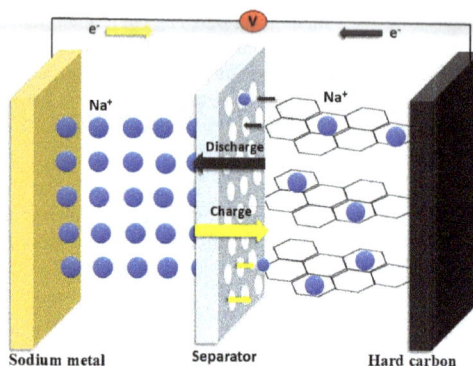

Figure 2. Schematic illustration of the working mechanism of SIBs with hard carbon anode. (Reproduced with permission from Ref. [17], Copyright 2025, Elsevier).

Potassium-Ion Batteries (PIBs) : PIBs operate on electrochemical principles similar to LIBs, but with the added benefit of faster ionic diffusion and enhanced rate performance. Different kinds of carbonaceous material are studied in the anode part of PIBs. Graphene offers a high surface area but suffers from restacking, which reduces the availability of active sites. Soft carbon, with its semi-graphitic nature, allows moderate ion diffusion. Biomass-derived hard carbon offers superior structural features, including expanded interlayer spacing and hierarchical porosity, which are ideal for accommodating the larger ionic radius of K^+ (1.38 Å), thus enhancing both ion transport and mechanical integrity[18]. Yuan et al. synthesized HC from chicken bone waste, resulting in a forest-like porous architecture that facilitates efficient ion transport and robust SEI formation. This material demonstrated an initial discharge of 470 mAh g^{-1} at 2C. It retained 250 mAh g^{-1} after 250 cycles at 0.2C, highlighting its potential for low-cost, scalable, high-performance PIB anode development[19].

Zinc-Based Batteries: Zinc–air batteries (ZABs) and zinc-ion batteries (ZIBs) offer advantages such as safety and cost-effectiveness, but are limited by issues including low conductivity and short cycle life. BCM has demonstrated significant potential in overcoming these limitations through structural and electrochemical enhancements. Zhao et al. synthesized N-doped porous carbon from soybean shells through pyrolysis, yielding a high-surface-area material with sulfonic groups and hierarchical porosity. This carbon exhibited excellent capacitive deionization performance, achieving an open-circuit voltage of 1.2 V, an adsorption capacity of 15.8 mg g^{-1}, and an average adsorption rate of 0.37 mg g^{-1} min^{-1} in a 40 mg L^{-1} NaCl solution, when powered by a ZAB system [20].

Li et al. developed a hydrogel electrolyte derived from hyaluronic acid, emphasizing its biocompatibility for application in ZIBs. The hydrophilic functional groups in the hydrogel effectively suppressed zinc anode corrosion and promoted uniform Zn nucleation. This electrolyte system achieved a high CE of 99.71% and maintained excellent cycling stability for over 5500 hours. Moreover, a Zn//LiMn$_2$O$_4$ pouch cell retained 82% of its capacity at a 3C rate after 1000 cycles, demonstrating the synergy between bio-derived electrolytes and zinc-based electrode systems [21]. Table 1 provides details of hard carbon materials derived from various biomass sources, along with their corresponding properties.

Table 1. Biomass-derived hard carbon for SIB.

Hard carbon from Biomass	Specific capacity (mAh/g)	Cycles	Current density (mA/g)	Initial Coulombic efficiency (%)	Ref.
Buckwheat	330	100	30	46%	[15]
Banaba	240	100	30	70.7%	[17]
Cedar wood bark	231	100	100	44%	[22]
Corncob	139	100	100	38%	[23]
Garlic peel	200	100	100	41%	[24]
Rambutan peel	225	100	30	55%	[25]

Biomass-derived Hard Carbon in Supercapacitors

Supercapacitors, also known as electrochemical capacitors, store energy through two primary mechanisms: electric double-layer capacitance (EDLC) and pseudocapacitance. EDLC is a non-Faradaic mechanism in which energy is stored by electrostatic charge separation at the electrode–electrolyte interfacial region, without electron transfer or faradaic reactions. This mechanism is highly reversible and relies on the adsorption of ions from the electrolyte onto the electrode surface. Carbon-based materials, including activated carbon (AC), carbon nanotubes (CNTs), graphene, carbon aerogels, and hard carbon, are commonly employed for EDLC electrodes due to their high surface area, good electrical conductivity, electrochemical stability, and economic viability [4,5]

Pseudocapacitance is a Faradaic process characterized by quick and reversible redox reactions that occur in the interfacial region of the electrode material. Materials exhibiting pseudocapacitive behavior include transition metal oxides, conductive polymers, and heteroatom-doped carbons. These materials can deliver higher energy densities than EDLCs but may suffer from reduced rate capabilities and shorter cycling life. Combining EDLC and pseudocapacitive mechanisms can enable hybrid capacitors with balanced energy and power characteristics.

Electric Double-Layer Capacitors (EDLCs): EDLCs store energy via non-Faradaic electrostatic charge accumulation at the electrode–electrolyte interface. Charge separation occurs at the interface between the electrode and electrolyte, forming an electric double layer. There is no charge transfer across the interface, and the process is highly reversible, resulting in extensive cycle life (up to 1 million cycles). Carbon-based materials, such as activated carbon (AC), carbon nanotubes (CNTs), graphene, carbon aerogels, and hard carbon, are widely used in EDLCs due to their high surface area, good conductivity, and chemical stability. Electrolytes in EDLCs include aqueous (e.g., H_2SO_4, KOH), organic (e.g., $TEABF_4$ in acetonitrile), or ionic liquids. EDLCs deliver high power density and rapid charge/discharge phenomena such as regenerative braking systems, memory backup, and UPS devices.

Pseudocapacitors: Pseudocapacitors operate through Faradaic redox reactions that occur at the electrode surface. Unlike EDLCs, the charge storage mechanism of Pseudocapacitors involves electron transfer between the electrode/electrolyte region and quick surface redox reactions. It offers higher specific capacitance and specific energy compared to EDLCs, although typically at a lower power density and reduced rate capability. Common electrode materials include transition metal oxides (e.g., MnO_2, RuO_2, $Ni(OH)_2$), conducting polymers (e.g., polyaniline, polypyrrole), and Heteroatom-doped carbon materials. Hence, these Pseudocapacitor materials are appropriate for moderate energy storage applications where high capacitance is prioritized.

Hybrid Supercapacitors (HSCs): Hybrid supercapacitors tailor the mechanisms of Faradaic and non-Faradaic reactions to attain maximum energy density as well as enhanced power density. They are categorized into:

> *Asymmetric Supercapacitors (ASCs):* One electrode behaves like a battery (Faradaic), and the other is a capacitor (non-Faradaic). Example: $AC//MnO_2$ or $AC//Ni(OH)_2$.

> *Battery-type HSCs:* Both electrodes undergo Faradaic reactions at different potentials or with distinct kinetics.

> *Composite HSCs:* Integrate capacitive and redox-active materials into a single electrode, e.g., carbon-metal oxide composites.

Hybrid systems are beneficial in portable electronics, EV start–stop systems, and grid support, where energy and power densities must be balanced. The efficiency and practicality of supercapacitors are evaluated using parameters such as specific capacitance ($F\ g^{-1}$), energy density ($Wh\ kg^{-1}$), power density ($W\ kg^{-1}$), cycle life, rate capability, and coulombic efficiency.

Biomass-Derived Hard Carbon for Hybrid Supercapacitors:

Biomass-derived carbon materials have garnered considerable attention in supercapacitor research due to their cost-effectiveness, renewability, and tunable porous architectures. Commonly referred to as biochars or activated carbons, these materials are typically produced via pyrolysis or carbonization of organic precursors such as banana stems, coconut shells, wood, and agricultural residues under inert atmospheres at elevated temperatures (800–1400 °C). Chemical activation using agents such as KOH or $ZnCl_2$ significantly enhances surface area and porosity, which are crucial for efficient ion transport and charge storage.

Thomas et al. synthesized biochar from banana stems modified with $FeCl_3/FeSO_4$ and polyaniline using the sonication method; the assembled system yields a specific capacitance of $315\ F\ g^{-1}$ [7]. Hard carbon (HC) possesses an amorphous, disordered, non-graphitizable arrangement and has surface functional groups that support the Faradic and non-Faradic reactions of the assembled cell. Tian et al. reported KOH-activated HC from *Linum usitatissimum*, with a specific capacitance of $434\ F\ g^{-1}$[8]. Similarly, Taniya et al. converted peanut shells into graphene-like nanosheets with a large porous surface area of $2070\ m^2\ g^{-1}$ and yields capacitance up to $186\ F\ g^{-1}$ [26].

Graphene-based materials have been engineered for flexible supercapacitor applications. For instance, Wang et al. developed a self-healing, stretchable fiber supercapacitor using graphene fibers, a gel electrolyte, and carboxylated polyurethane [27]. Sevilla et al. achieved hydrochar from biomass precursors with surface areas up to $2700\ m^2\ g^{-1}$ [28]. Hard carbon (HC), which primarily contributes via EDLC, can also exhibit pseudocapacitance due to its surface functionalities. Jiang

Electrocatalysts and Advanced Materials for Sustainable Energy Storage   Materials Research Forum LLC
Materials Research Foundations 182 (2025) 87-99   https://doi.org/10.21741/9781644903797-7

et al. reported an AC/HC hybrid capacitor with 21.5 F g^{-1} and 22.6 Wh kg^{-1} at 480 W kg^{-1}[29]. Richey et al. demonstrated that onion-like carbons offer stable ion dynamics under high currents, explained by the mechanism of *in situ* infrared spectro-electrochemical analysis [30]. Zhang et al. investigated the use of lotus stems and leaves which are biomass sources naturally rich in intrinsic potassium as precursors for producing porous carbon materials. Upon pyrolysis at 800 °C for 3 hours, the resulting carbons exhibited high specific surface areas of 1610 m^2 g^{-1} for stems and 1039 m^2 g^{-1} for leaves. The inherent potassium acted as a self-activating agent, promoting pore formation and enhancing surface area, making these materials highly suitable for energy storage applications such as supercapacitors [31].

Fengbin et al. studied the naturally nitrogen- and oxygen-doped hard carbon (NOHC) and analyzed its electrochemical properties for hybrid potassium-ion capacitors (KIHCs). The assembled cell consists of an NOHC anode and an activated carbon cathode, delivering an energy of 68 Wh kg^{-1} at a power of 2000 W kg^{-1}. The device exhibits 80% cyclic retention up to 1000 cycles at a current density of 1 A/g^{-1}, which ensures the material is best suited for an energy storage system [32]. In a complementary study, N. Guo et al. explored the use of cost-effective porous carbon synthesized from soybean roots as an electrode material, utilizing EMIM BF$_4$ ionic liquid and a 6 M KOH electrolyte for supercapacitors. The assembled device exhibited a specific capacitance of 276 F g^{-1} at 0.5 A g^{-1}, in 6 M KOH and maintained cycling durability of 98% up to 10,000 charge–discharge cycles at 5 A g^{-1}. On the other hand, in EMIM BF$_4$ ionic liquid delivers the energy of 100.5 Wh kg^{-1} at a power density of 4353 W kg^{-1} which underscores the outstanding resilience of the biomass-derived material [33].

These findings highlight the promise of sustainable, biomass-derived carbon architectures in next-generation supercapacitor and hybrid capacitor technologies. **Table 2** summarizes the electrochemical parameters of various biomass-derived carbon materials reported in recent studies. In hybrid supercapacitors (HSCs), biomass-derived carbon materials play a dual role: the carbon coating facilitates fast electron transport, while the underlying active material contributes to pseudocapacitance through redox reactions. This synergistic mechanism results in enhanced energy/power densities, high-rate capability, and exceptional cyclic durability. Their eco-friendly nature, tunable properties, and structural versatility make them ideal candidates for future-generation flexible, solid-state, and high-performance supercapacitors.

Table 2. Hard Carbon as Electrode Material for Supercapacitors

Hard carbon from Biomass	Specific capacitance F g^{-1}	Energy density Wh kg^{-1}	Power density W kg^{-1}	No. of cycle	Ref.
Popcorn	286	53		10 000	[34]
Dry elm samara	310	25.4	15kW	5000	[35]
Wheat Straws	275	-	-	10,000	[36]
Cellulose	160	17.7	180	10000	[37]
Lignin	123	5	15	5000	[38]

Conclusion

Hard carbon is a promising electrode material for batteries and supercapacitors due to its sustainability, scalability, and electrochemical performance. However, challenges such as low initial coulombic efficiency and irreversible capacity loss remain significant barriers to practical implementation. Future research should focus on surface engineering, heteroatom doping, and structural optimization to enhance ion accessibility, improve reversibility, and stabilize electrode–electrolyte interfaces. Advancements in scalable synthesis and device-level integration will be key to transitioning hard carbon into commercial energy storage technologies.

References

[1] H. Pan, Y.S. Hu, L. Chen, Room-temperature stationary sodium-ion batteries for large-scale electric energy storage, Energy Environ Sci 6 (2013) 2338–2360. https://doi.org/10.1039/c3ee40847g.

[2] X. Zhu, Y. Zeng, X. Zhao, D. Liu, W. Lei, S. Lu, Biomass-Derived Carbon and Their Composites for Supercapacitor Applications: Sources, Functions, and Mechanisms, EcoEnergy (2025). https://doi.org/10.1002/ece2.70000.

[3] P. Sohtun, D. Deb, N. Bora, R. Goswami, P.K. Choudhury, R. Boddula, P.K. Sarangi, R. Kataki, T.A. Kurniawan, Agriculture biomass-derived carbon materials for their application in sustainable energy storage, Carbon Letters (2025). https://doi.org/10.1007/s42823-025-00884-9.

[4] Aravindan, V., Ulaganathan, M. and Madhavi, S., 2016. Research progress in Na-ion capacitors. *Journal of Materials Chemistry A*, 4(20), pp.7538-7548. https://doi.org/10.1039/C6TA02478E

[5] Jia, R., Shen, G. and Chen, D., 2020. Recent progress and future prospects of sodium-ion capacitors. *Science China Materials*, 63(2), pp.185-206. https://doi.org/10.1007/s40843-019-1188-x.

[6] X. Dou, I. Hasa, D. Saurel, C. Vaalma, L. Wu, D. Buchholz, D. Bresser, S. Komaba, S. Passerini, Hard carbons for sodium-ion batteries: Structure, analysis, sustainability, and electrochemistry, Materials Today 23 (2019) 87–104. https://doi.org/10.1016/j.mattod.2018.12.040.

[7] D. Thomas, N.B. Fernandez, M.D. Mullassery, R. Surya, Iron oxide loaded biochar/polyaniline nanocomposite: synthesis, characterization and electrochemical analysis, Inorg. Chem. Commun. 119 (2020), https://doi.org/ 10.1016/j.inoche.2020.108097

[8] W.Tian, P. Ren, X. Hou, Z. Guo, R. Xue, Z. Chen, Y. Jin, Biomass derived N/O self doped porous carbon for advanced supercapacitor electrodes, Ind. Crops Prod. 202 (2023), https://doi.org/10.1016/j.indcrop.2023.117032.

[9] Y.P. Wu, C. Jiang, C. Wan, R. Holze, Modified natural graphite as anode material for lithium ion batteries, n.d.

[10] M. Tariq, K. Ahmed, Z. Khan, M.P. Sk, Biomass-Derived Carbon Dots: Sustainable Solutions for Advanced Energy Storage Applications, Chem Asian J (2025). https://doi.org/10.1002/asia.202500094.

[11] W. ur Rehman, Y. Ma, Z. khan, F.Z.A. Laaskri, J. Xu, U. Farooq, A. Ghani, H. Rehman, Y. Xu, Biomass-derived carbon materials for batteries: Navigating challenges, structural diversities, and future perspective, Next Materials 7 (2025). https://doi.org/10.1016/j.nxmate.2024.100450.

[12] Y. Zhang, Y. Wang, Y. Meng, G. Tan, Y. Guo, D. Xiao, Porous nitrogen-doped carbon tubes derived from reed catkins as a high-performance anode for lithium ion batteries, RSC Adv 6 (2016) 98434–98439. https://doi.org/10.1039/c6ra21620j.

[13] L. Xie, C. Tang, Z. Bi, M. Song, Y. Fan, C. Yan, X. Li, F. Su, Q. Zhang, C. Chen, Hard Carbon Anodes for Next-Generation Li-Ion Batteries: Review and Perspective, Adv Energy Mater 11 (2021). https://doi.org/10.1002/aenm.202101650.

[14] M.C. Sun, S.L. Qi, Y.H. Zhao, C.X. Chen, L.C. Tan, Z.L. Hu, X.L. Wu, W.L. Zhang, Advances in the use of biomass-derived carbons for sodium-ion batteries, Xinxing Tan Cailiao/New Carbon Materials 40 (2025) 1–49. https://doi.org/10.1016/S1872-5805(25)60953-X.

[15] U. Kydyrbayeva, Y. Baltash, O. Mukhan, A. Nurpeissova, S.S. Kim, Z. Bakenov, A. Mukanova, The buckwheat-derived hard carbon as an anode material for sodium-ion energy storage system, J Energy Storage 96 (2024). https://doi.org/10.1016/j.est.2024.112629.

[16] C. Yu, H. Hou, X. Liu, Y. Yao, Q. Liao, Z. Dai, D. Li, Old-loofah-derived hard carbon for long cyclicity anode in sodium ion battery, Int J Hydrogen Energy 43 (2018) 3253–3260. https://doi.org/10.1016/j.ijhydene.2017.12.151.

[17] P. Keerthana, R. Sivasubramanian, N. Priyadharsini, Electrochemical studies on Lagerstroemia speciosa based hard carbon anode for sodium ion batteries, Inorg Chem Commun 178 (2025). https://doi.org/10.1016/j.inoche.2025.114417.

[18] Z. Liu, J. Chen, K. He, Z. Fang, H. Zhang, Z. Gong, S. Ding, T. Wang, G. Zhang, Q. Liu, From laboratory to mass production: mechanistic insights and optimization of eco-friendly carbon-based anodes from biomass for potassium-ion batteries, Energy Materials 5 (2025). https://doi.org/10.20517/energymater.2024.154.

[19] X. Yuan, B. Zhu, J. Feng, C. Wang, X. Cai, R. Qin, Biomass bone-derived, N/P-doped hierarchical hard carbon for high-energy potassium-ion batteries, Mater Res Bull 139 (2021). https://doi.org/10.1016/j.materresbull.2021.111282.

[20] C. Zhao, G. Liu, N. Sun, X. Zhang, G. Wang, Y. Zhang, H. Zhang, H. Zhao, Biomass-derived N-doped porous carbon as electrode materials for Zn-air battery powered capacitive deionization, Chemical Engineering Journal 334 (2018) 1270–1280. https://doi.org/10.1016/j.cej.2017.11.069.

[21] G. Li, Z. Zhao, S. Zhang, L. Sun, M. Li, J.A. Yuwono, J. Mao, J. Hao, J. (Pimm) Vongsvivut, L. Xing, C.X. Zhao, Z. Guo, A biocompatible electrolyte enables highly reversible Zn anode for zinc ion battery, Nat Commun 14 (2023). https://doi.org/10.1038/s41467-023-42333-z.

[22] M. Lu, Y. Huang, C. Chen, Cedarwood Bark-Derived Hard Carbon as an Anode for High-Performance Sodium-Ion Batteries, Energy and Fuels 34 (2020) 11489–11497. https://doi.org/10.1021/acs.energyfuels.0c01841.

[23] Q. Jiang, Z. Zhang, S. Yin, Z. Guo, S. Wang, C. Feng, Biomass carbon micro/nano-structures derived from ramie fibers and corncobs as anode materials for lithium-ion and sodium-ion batteries, Appl Surf Sci 379 (2016) 73–82. https://doi.org/10.1016/j.apsusc.2016.03.204.

[24] V. Selvamani, R. Ravikumar, V. Suryanarayanan, D. Velayutham, S. Gopukumar, Garlic peel derived high capacity hierarchical N-doped porous carbon anode for sodium/lithium ion cell, Electrochim Acta 190 (2016) 337–345. https://doi.org/10.1016/j.electacta.2016.01.006.

[25] A.A. Arie, H. Kristianto, H. Muljana, L. Stievano, Rambutan peel based hard carbons as anode materials for sodium ion battery, Fullerenes Nanotubes and Carbon Nanostructures 27 (2019) 953–960. https://doi.org/10.1080/1536383X.2019.1671372.

[26] Purkait T, Singh G, Singh M, Kumar D, Dey RS (2017) Large area few-layer graphene with scalable preparation from waste biomass for high-performance supercapacitor. Sci Rep 7(1):15239. https://doi.org/10.1038/s41598-017-1 5463-w

[27] S. Wang, N. Liu, J. Su, L. Li, F. Long, Z. Zou, X. Jiang, Y. Gao, Highly Stretchable and Self-Healable Supercapacitor with Reduced Graphene Oxide Based Fiber Springs, ACS Nano 11 (2) (2017) 2066–2074.https://doi.org/10.1021/acsnano.6b08262

[28] Sevilla, M., Fuertes, A.B. and Mokaya, R., 2011. High density hydrogen storage in superactivated carbons from hydrothermally carbonized renewable organic materials. Energy & Environmental Science, 4(4), pp.1400-1410. https://doi.org/10.1039/C0EE00347F

[29] Jiangfeng Nia, Youyuan Huangb, Lijun Gao, A high-performance hard carbon for Li-ion batteries and supercapacitors application, Journal of Power Sources 223 (2013) 306e311, http://dx.doi.org/10.1016/j.jpowsour.2012.09.047.

[30] F.W. Richey, B. Dyatkin, Y. Gogotsi, Y.A. Elabd, Ion dynamics in porous carbon electrodes in supercapacitors using in situ infrared spectroelectrochemistry, J. Am. Chem. Soc. 135 (34) (2013) 12818–12826.

[31] Y. Zhang, S. Liu, X. Zheng, X. Wang, Y. Xu, H. Tang, F. Kang, Q.H. Yang, J. Luo, Biomass organs control the porosity of their pyrolyzed carbon, Adv. Funct. Mater. 27 (2017) 1–8, https://doi.org/10.1002/adfm.201604687.

[32] Huang, F., Liu, W., Wang, Q., Wang, F., Yao, Q., Yan, D., Xu, H., Xia, B.Y. and Deng, J., 2020. Natural N/O-doped hard carbon for high performance K-ion hybrid capacitors. *Electrochimica Acta, 354*, p.136701. https://doi.org/10.1016/j.electacta.2020.136701

[33] Guo, N., Li, M., Wang, Y., Sun, X., Wang, F. and Yang, R., 2016. Soybean root-derived hierarchical porous carbon as electrode material for high-performance supercapacitors in ionic liquids. *ACS applied materials & interfaces, 8*(49), pp.33626-33634. https://doi.org/10.1021/acsami.6b11162

[34] Hou, J., Jiang, K., Wei, R., Tahir, M., Wu, X., Shen, M., Wang, X. and Cao, C., 2017. Popcorn-derived porous carbon flakes with an ultrahigh specific surface area for superior performance supercapacitors. *ACS applied materials & interfaces, 9*(36), pp.30626-30634. https://doi.org/10.1021/acsami.7b07746

[35] Chen, C., Yu, D., Zhao, G., Du, B., Tang, W., Sun, L., Sun, Y., Besenbacher, F. and Yu, M., 2016. Three-dimensional scaffolding framework of porous carbon nanosheets derived from plant wastes for high-performance supercapacitors. *Nano Energy*, *27*, pp.377-389. https://doi.org/10.1016/j.nanoen.2016.07.020

[36] Liu, W., Mei, J., Liu, G., Kou, Q., Yi, T. and Xiao, S., 2018. Nitrogen-doped hierarchical porous carbon from wheat straw for supercapacitors. *ACS Sustainable Chemistry & Engineering*, *6*(9), pp.11595-11605. https://doi.org/10.1021/acssuschemeng.8b01798

[37] Tian, X., Zhu, S., Peng, J., Zuo, Y., Wang, G., Guo, X., Zhao, N., Ma, Y. and Ma, L., 2017. Synthesis of micro-and meso-porous carbon derived from cellulose as an electrode material for supercapacitors. *Electrochimica Acta*, *241*, pp.170-178. https://doi.org/10.1016/j.electacta.2017.04.038

[38] Zhang, W., Lin, H., Lin, Z., Yin, J., Lu, H., Liu, D. and Zhao, M., 2015. 3 D hierarchical porous carbon for supercapacitors prepared from lignin through a facile template-free method. *ChemSusChem*, *8*(12), pp.2114-2122. https://doi.org/10.1002/cssc.201403486

Electrocatalysts and Advanced Materials for Sustainable Energy Storage   Materials Research Forum LLC
Materials Research Foundations 182 (2025) 100-117   https://doi.org/10.21741/9781644903797-8

Chapter 8

Transition Metal Oxide Catalysts as Supercapacitor Electrode Materials for Sustainable Energy Storage Applications

AVILA Josephine B.[1,a*], Mary TERESITA V.[1,b*], MATHUMITHA Babu[1,c]

Department of Chemistry, Stella Maris College (Autonomous), Chennai – 600 086, India

[a]avilajosephine@stellamariscollege.edu.in, [b]maryteresita@stellamariscollege.edu.in, [c]mathu2k01@gmail.com

Abstract

In recent decades, the growing dependency on non-renewable fossil fuels has presented a significant risk to human sustainability, emphasizing the critical need for developing non-polluting and economically achievable energy storage technologies. Among these, supercapacitors have accumulated significant attention for their vital role in energy conservation, offering distinct advantages such as compact form factor, lightweight construction, elevated power density, and extended operational lifespan. They have garnered global attention for their unique properties and potential applications. Several studies have been devoted to improving electrode materials, with particular emphasis on transition metal oxide (TMO)-based systems. These materials are in high demand for their high theoretical pseudocapacitance, making them ideal candidates for fabricating advanced composite electrodes in supercapacitor applications. Enhancing these electrodes is critical role for achieving superior energy density, greater specific power, and rapid charge–discharge capabilities, the factors that collectively optimize the performance and efficiency of supercapacitors. This chapter presents a comprehensive overview of the latest developments in TMO composite materials. It also focuses on the fabrication techniques, structural characterization, electrochemical behavior, and the evaluation of various transition metal oxide (TMO)-based electrode materials in supercapacitor systems. By systematically examining these features, the chapter aims to unveil the full potential of TMO composites for energy storage applications and to advance their continued evolution within the field.

Keywords

Energy Storage, Pseudo Capacitance, Transition Metal Oxides, Composite Electrodes

Contents

Electrocatalysts and Advanced Materials for Sustainable Energy Storage   Materials Research Forum LLC
Materials Research Foundations 182 (2025) 100-117   https://doi.org/10.21741/9781644903797-8

Introduction

The need to reduce the use of fossil fuels has mandated the prioritization of renewable energy sources, which, in turn, has focused attention on energy storage devices (ESDs) that support sustainable energy systems. ESDs, including batteries, capacitors, and supercapacitors (SCs), have gained commercial feasibility due to their advantageous energy and power densities, superior capacity retention, and cost-effectiveness. SCs are more efficient than lithium-ion batteries (LIIBs) due to their superior specific power, cost-effectiveness, diverse properties, scalability, long-term durability, and electrochemical stability [1]. The composition of electrode materials plays an important role in the change of its charge storage mechanism and electron transfer dynamics, which in turn play a key role in increasing its overall performance, improving its efficiency, and expanding its application potential. Supercapacitors (SCs) can store electrical charge through a pseudo-capacitive mechanism or through electrical polarization in the electrolyte-electrode interface [2,3]. On the whole, the supercapacitors (SCs) exhibit valuable practical applications across electric vehicles (EVs), portable electronic gadgets, and backup power systems, where rapid energy delivery and reliability are essential.

Electrochemical capacitors, including hybrid SCs, pseudocapacitors, and electric double-layered capacitors [4] have been identified using various methods. The development of hybrid supercapacitors (SCs) and pseudocapacitors depends on the specialized formulation in the synthesis of metal oxides (MOs), along with precise control over their morphology, crystalline structure, expected capacitance, oxidation states, and elemental composition. MOs like MnO_2, Co_3O_4, RuO_2, and V_2O_5 are potential substitutes for SC electrodes due to their pseudo-capacitive properties, high capacitance, and compatibility with carbon-based materials [5]. Electrical Double Layer Capacitors (EDLCs,) have higher charge-discharge cycles than pseudocapacitors, and have a lower specific capacitance (C_{sp}) [6]. SCs are generally classified into symmetric and asymmetric configurations, based on the composition and pairing of their electrode materials. Symmetric SCs exhibit uniform properties, while pseudocapacitors have increased C_{sp} and enhanced reversible behaviour due to favourable redox processes [5-7]. Various pseudocapacitive materials, including TMOs like FeO_x/CNF [8], SiO_2–Cu [9], and SiO_2–ZnO [10], have been extensively studied and discussed in scientific literature. EDLCs are generally carbon-based materials such as activated

Electrocatalysts and Advanced Materials for Sustainable Energy Storage   Materials Research Forum LLC
Materials Research Foundations 182 (2025) 100-117   https://doi.org/10.21741/9781644903797-8

carbon, graphitic carbon, carbon nanotubes, graphene nanofibers, reduced graphene oxide, and carbon black, due to their high surface area and excellent conductivity.

Pure transition metal compounds (TMCs) and their composites are crucial for energy storage due to their ability to create links with other materials and their ability to change valence states [11]. The valence states of transition metals within these composites play a crucial role in influencing their performance in solid-state supercapacitors (SCs). High valence states contribute to improved electronic conductivity, facilitating efficient charge transport throughout electrochemical processes. The charge storage is also affected when there are differences in valence states of the metals. Some specific valence states produce large active sites for charge storage, which improves the overall capacitance [12-13]. Electrochemical performance is delicately tied to redox interactions involving transition metal ions, where variations in valence states directly influence the reversible storage of electrical energy. Careful regulation and understanding of these valence dynamics are important for maintaining structural integrity over extended charge–discharge cycling. Consequently, researchers are increasingly concentrating on modifying the valence states of supercapacitor (SC) electrode materials through advanced synthesis techniques to optimize device efficiency and durability.

Knowing the relationship between valence states and electrochemical performance is critical for developing high-performance ESDs and efficient SC technologies [14-16]. On analysizing the electrochemical and physicochemical properties of transition metal oxides (TMOs) , it has been found that they have greatly accelerated the development of high-efficiency supercapacitors (SCs). This enhanced performance is largely attributed to the flexible oxidation states of these metals, which directly influence their charge storage capabilities. Metal oxides exhibiting strong supercapacitive behavior serve as promising substrates for fabricating electrode materials that are capable of delivering high specific capacitance (Csp). Ruthenium oxide (RuO_2) is extensively studied due to its robust ionic conductivity, prolonged lifespan, high C_{sp}, exceptional thermal stability [17,18], and reversible redox processes. Its affordable cost, large range of applications, and impressive estimated C_{sp} of around 1370 Fg^{-1} [19] make it a popular choice for energy storage applications, such as supercapacitors and electrodes for lithium batteries.

Integrating metals or their oxides into composite structures significantly improves electrochemical characteristics, including enhanced ionic permeability, abundant electroactive sites, versatile valence states, superior structural stability, and enhanced electrical conductivity. Carbon materials are used in novel electrode structures, and studying energy storage mechanisms in TMO is crucial for boosting the efficiency of SC electrodes [20-22]. Common TMO electrodes include Bi_2O_3 [23], MnO_2 [24], NiO [25], Mn_3O_4 [26], RuO_2 [27], and Co_3O_4 [28], both in symmetric and asymmetric configurations. RuO_2 exhibits a comparatively lower Csp than other transition metal oxides such as $NiOx$, $CoOx$, and MnO_2. This chapter offers a comprehensive overview of TMO composites as potential electrode materials for SC applications, with particular emphasis on synthetic strategies and recent advancements in fabrication methodologies employed in their development.

Development of Transition Metal Oxides for Supercapacitors

The Supercapacitors took over seventy years to be economically viable. The Leyden Jar was the original form of the capacitor. The glass vessel's body served as the electrode for the thin metallic foil that made up the dielectric jar. In 1957, General Electric established the first SC, as stated by Ragha Vendra et al. [29]. The non-Faradaic charge storage mechanism—characterized by the constant electric charge maintained between the electrode and electrolyte—serves as the foundational principle behind EDLCs. However, to elevate the performance of SCs, the development of new electrode materials is essential. Extensive research has explored a variety of electrochemically active compounds to boost their efficiency and energy storage capabilities. Since the 1980s, RuO_2 has garnered considerable attention as a promising pseudocapacitor material, offering insights into charge storage mechanisms governed by redox reactions, ion intercalation, and Faradaic electro-sorption processes. Shortly after, Li-ion capacitors, also known as hybrid SCs, were created in 2007 by FDK-Japan. Later, in 2014, Jin et al. [30] from George Washington University developed a cost-effective, high-performing ultra-capacitor using graphene and CNTs. It was designed utilizing the rod rolling method and ink made of carbon NMs.

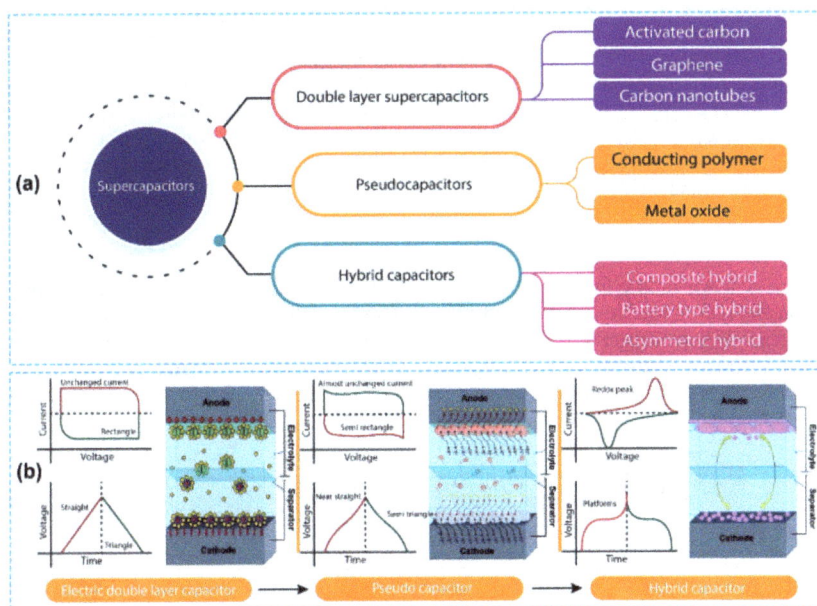

Figure 1. (a) Types of SCs and (b) Associated operational mechanisms (Reproduced with permission from Ref. [4], Copyright 2024, Elsevier)

In 2015, the Agency for Research, Science, and Technology in Singapore developed an asymmetric SC using metal nitride and graphene as core components. The research team employed atomic layer deposition to fabricate two distinct materials directly onto vertically aligned graphene nanosheets, enhancing the structural integration and functional performance of the device. These capacitors exhibited excellent power density and were very capacitive [31]. Subsequently, researchers at the VTT Technical Research Centre of Finland engineered a hybrid nano-electrode structure featuring porous silicon coated with titanium nitride. This novel design aims to optimize the complementary advantages of EDLCs and pseudocapacitors, while offsetting their drawbacks, thereby enhancing the overall performance of hybrid SCs. The SC global market was estimated to be valued at $887 million in 2020 [5]. TMO-based redox-active materials are the most often utilized for SC electrodes. These substances are commonly used in hybrid SCs and exist in both hydroxide and oxide forms [9].

The Basis of Supercapacitors

Supercapacitors (SCs), also known as ultracapacitors or electrochemical capacitors, they produce higher capacitance than conventional capacitors due to their high-surface-area electrodes and thin separators. Supercapacitors are generally classified into three main categories: EDLCs, pseudocapacitors, and hybrid capacitors. EDLCs operate by storing charge electrostatically at the electrode–electrolyte interface, without engaging in Faradaic processes. This non-Faradaic mechanism grants EDLCs remarkable cycle stability, particularly when fabricated using porous carbon-based materials such as graphene and carbon nanotubes. In contrast, pseudocapacitors rely on rapid Faradaic redox reactions involving conducting polymers or metal oxides/hydroxides, offering higher energy but lower power density and structural stability. Hybrid capacitors combine EDLC and pseudocapacitive electrodes to balance the energy and power performance, with lithium-ion capacitors which is being a prominent example. Figure 1 shows the different types of supercapacitors and their mechanisms [4].

Materials for Supercapacitors

Supercapacitors are made from various materials, including carbon, metal oxides like iron, cobalt, manganese, nickel, ruthenium, and conducting polymers. Carbon nanotubes (CNTs) are gaining prominence in supercapacitor technology owing to their exceptionally high surface area, which facilitates efficient charge storage. Additionally, composite materials—engineered by integrating multiple constituents—are being actively explored to further enhance the electrochemical performance and stability of supercapacitor devices.

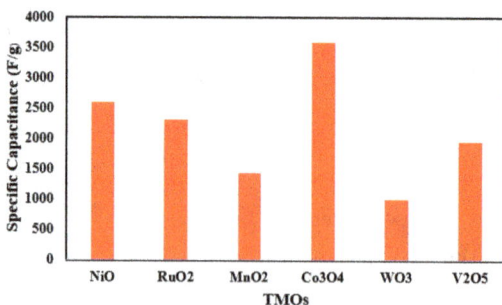

Figure 2. Theoretical specific capacitance of some TMOs (Reproduced with permission from Ref. [34], Copyright 2022, Intech Open)

Nanostructured materials with enhanced capacitive performance characteristics and more surface area are being considered for supercapacitor electrodes. The fabrication of high-performance supercapacitor devices primarily involves the fabrication of nanoengineered materials [32, 33]. Figure 2 depicts the theoretical specific capacitance of some of TMOs [34].

Metal Oxides/Hydroxides Composite Electrodes

Metal oxides offer higher energy density and better electrochemical stability for supercapacitors than conventional carbon materials. They store energy and exhibit electrochemical Faradaic reactions. This review explores the progress in developing metal oxides and hydroxides—including ruthenium oxide, manganese oxide, nickel oxide, iron oxide, cobalt oxide, nickel hydroxide, cobalt hydroxide, and nickel cobaltite—as potential supercapacitive materials, with a particular focus on synthesis via microwave-assisted techniques.

(i) Ruthenium oxides (RuO₂): Ruthenium oxides (RuO_2) offer stability, large Faradaic activity, and ion adsorption pseudo capacitance [35], but are not commercially viable due to their high cost. Carbon nanotubes (CNTs) are attractive electrode materials for supercapacitors due to their useful space and accessibility [36]. The hybridization of CNTs with RuO_2 has been extensively investigated for supercapacitor applications. RuO_2/multi-walled carbon nanotube (MWCNT) nanocomposites were synthesized using a microwave-assisted technique, employing ruthenium $RuCl_3$ solution as the precursor and $NH_3 \cdot H_2O$ as the precipitating agent. The process involved removing chlorine ions and $NH_3.H_2O$, and $Ru(OH)_3$ transformed into RuO_2 nanoparticles without extra annealing.[37] RuO_2 nanoparticles were attached to MWCNTs' sidewalls and inner cavities, enhancing their C_{sp}. The RuO_2/MWCNTs electrode demonstrated superior specific capacitance (Csp) compared to other electrode configurations, primarily attributed to the enhanced contribution of RuO_2 nanoparticles. Specifically, the MWCNTs composite containing 40% RuO_2 exhibited a capacitance of 493.9 $F \cdot g^{-1}$, markedly higher than the 232.5 $F \cdot g^{-1}$ recorded for the composite with 20% RuO_2. The composites may have promising applications in high charge storage capacity devices [38-40].

(ii) Manganese oxide: Manganese oxides, including MnO_2, Mn_2O_3, Mn_3O_4, and MnO, have different forms due to their oxidation states. Mn_3O_4 is a stable mixed oxide with a spinel structure, widely used in supercapacitors, lithium batteries, fuel cells, and metal-air batteries [41-44]. It also acts as a catalyst for waste gas decomposition, making it suitable for controlling air pollution. Crystallized Manganese oxide (MnO_2) has several crystalline structures, including α-, β-, γ-, and δ-MnO_2 [45]. MnO_2 is regarded as a cost-effective transition metal oxide for supercapacitor applications, owing to its low production cost, environmental benignity, and high theoretical Csp. However, its use often produces low C_{sp} due to poor electrical conductivity and lack of accessible surface area [46,47]. MnO_2's superior electrochemical performance depends on factors such as structure, particle size, surface morphology, homogeneity, and bulk density [48-50]. Variations in synthesis approaches result in distinct physical and chemical characteristics, influencing key parameters such as crystallinity, morphological features, specific surface area, and cycling stability. Numerous routes and techniques for preparing MnO_2 have been developed, but most require extensive mechanical mixing, long duration, high temperature, and energy-wasting [51-55].

(iii) Nickel oxide: In recent years, nanostructured nickel oxide (NiO) has attracted substantial interest as an economical and highly promising electrode material for supercapacitors, owing to its elevated theoretical capacitance and favorable electrochemical characteristics. Nonetheless, practical implementation is hindered by its inherently low electrical conductivity and restricted

surface area—particularly evident in commercial β-nickel hydroxide [β-Ni(OH)$_2$]—which limits its overall performance. To address this, researchers have explored various synthesis techniques such as Chemical Bath Deposition, Potentiostatic Anodization, and Hydrothermal methods [56], with automatic spray pyrolysis emerging as a reliable process for producing uniform NiO thin films. This technique involves the thermal decomposition of fine droplets over heated substrates to create high-porosity films. The choice of nickel precursor (nitrate, acetate, sulfate) significantly influences the morphology and properties of NiO, as demonstrated by Deepa M. Audi's findings—where nickel chloride produced a mud-like structure while nickel nitrate yielded spherical grains. In the study of the spray pyrolysis technique [57, 58] using different nickel sources to synthesize NiO thin film electrodes, which were then analyzed using XRD, contact angle measurements, and SEM, followed by electrochemical evaluation via CH instrumentation.

Synthesis Techniques for Transition Metal Oxides

TMO nanoparticles and other chemicals for TMO composites can be made using synthesis methods such as hydrothermal, sonochemical, co-precipitation, solvothermal, and sol-gel. Each process has particular requirements depending on the requirements of the sample [59,60]. Figure 3 gives the various methods used for the synthesis of nanoparticles [4].

Top-down and bottom-up method: Nanoparticles (NPs) are synthesized on wafers using top-down and bottom-up methods, but both processes are expensive and challenging to control. Although bottom-up methods are less expensive and have the ability to regulate deposit regions ranging from 1 to 20 nm, they only yield small amounts and need to be purified. Hydrothermal synthesis is necessary for precisely controlled NP development in high-temperature and high-pressure aqueous conditions. They are appropriate candidates for ESDs because they permit exact size, shape, and crystallinity. After washing and annealing, ZnMn$_2$O$_4$ NPs were produced by SM. Courtel and his group via hydrothermal synthesis [60].

Sonochemical method: Sonochemical methods produce small, regulated nanoparticles (NPs) with distinct properties by using high-frequency sound waves to form NPs in liquids. Higher yields and improved selectivity are produced by these methods, which also save energy and the environment [61]. Chen and his group produced ZrO$_2$ and ZrO$_2$ chitosan, a compound that exhibited advantageous electrochemical properties. An ultrasonicator aids in the production of the precursor solution, and ZrO$_2$ precipitates are created and reduced to a fine powder following 30 minutes of exposure. For ZrO$_2$ chitosan composites, comparable procedures can be applied [62].

Co-precipitation method: Co-precipitation's capacity to precipitate multiple components and produce NPs at the same time makes it special. Because of its versatility and ability to manipulate the characteristics and composition of NPs, it is essential for creating materials with specific uses and enhanced functionality in contemporary electronics. Its scalability and convenience of application further increase its relevance in the synthesis of NPs [63]. A neutral and homogeneous precursor solution was initially prepared for the synthesis of cerium oxide nanoparticles (CeO$_2$ NPs). Subsequent centrifugation of this solution promotes the formation of CeO$_2$ precipitates. These precipitates were then cleaned and dried to yield NPs (with a size of about 30 nm) [64].

Solvothermal and Sol-gel method: The solvothermal method is a adaptable method for preparing nanomaterials, enabling the formation of high-purity, narrow-sized nanoparticles (NPs) suitable for sensors [65], ESDs, and nanoelectronics [66]. It allows for the modification   of NP characteristics and regulation of reaction parameters, making it an effective tool in material science and nanotechnology. On the contrary the sol-gel method, offers good control over size, shape, and composition, making it essential for producing NPs with specific qualities like stability, controlled release, and distinctive electrochemical properties [67 -69].

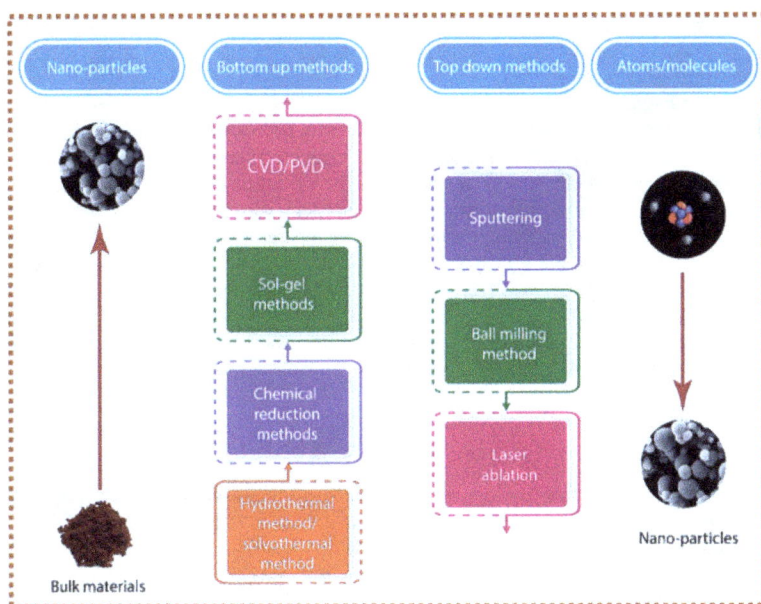

Figure 3. Schematic representation of the Top-down and Bottom-up approaches (Reproduced with permission from Ref. [4], Copyright 2024, Elsevier)

Hydrothermal method: It is the most widely used techniques for fabricating nanomaterials, applying a solution-phase reaction mechanism. This method enables the formation of nanostructures across a broad temperature spectrum, ranging from ambient conditions to elevated thermal environments. To control the morphology of the materials to be prepared, either low-pressure or high-pressure conditions can be used, depending on the vapor pressure of the main composition in the reaction [70]. A direct hydrothermal method was used to synthesize a spherical shaped NiO powders. In this method organic surfactants were used as structural templates and urea as a hydrolysis-regulating agent. Different types of surfactants like cationic (cetyl trimethyl ammonium bromide), anionic (sodium dodecyl sulfate), and nonionic (Triton X-100) were used s systematically and examined to observe the changes in parameters such as surface area, pore size, pore volume, and electrochemical behavior of the resulting NiO powders. Electrochemical analyses showed that the charge-storage mechanism in these NiO-based electrodes is predominantly Faradaic, rather than purely capacitive. The effect on the capacitance was clearly observed when ionic nature surfactant were used in the preparation of NiO powders [56].

Electrocatalysts and Advanced Materials for Sustainable Energy Storage   Materials Research Forum LLC
Materials Research Foundations 182 (2025) 100-117   https://doi.org/10.21741/9781644903797-8

Evaluating Synthesis Strategies: Benefits vs. Challenges

Hydrothermal synthesis is a solution-based technique conducted under elevated temperature and pressure conditions, enabling the production of high-purity, well-crystallized materials with precisely controlled physicochemical properties. Co-precipitation is simple and cost-effective, ideal for large-scale production, but has limited particle features [71-74]. Sonochemical synthesis leverages ultrasonic irradiation to efficiently control particle size and morphology, offering a rapid and energy-effective route for nanomaterial production. Solvothermal synthesis, conducted in non-aqueous solvents under elevated temperature and pressure, provides exceptional purity and precise control over structural features, while accommodating a diverse array of precursor materials. The sol-gel technique is versatile and works at low-temperature processing, but can contain potentially dangerous solvents [75]. The physicochemical characteristics—namely size, shape, and distribution—of nanoparticles (NPs) are governed by synthesis parameters such as pressure, temperature, solution molarity, and precursor pH. Sonochemical synthesis modulates nanoparticle agglomeration and morphology through variables including ultrasonic wave velocity, wavelength, frequency, temperature, applied pressure, and cooling rate. Sol–gel processing enables the fabrication of diverse forms, including monoliths, thin films, fibers, and submicron powders, across a wide range of material systems. Co-precipitation techniques rely on temperature, pressure, pH, precursor concentration, seed material, solvent type, and reaction duration to fine-tune the morphology and dimensional attributes of the resulting nanoparticles. The size ranges from 16 nm to 80 nm [76, 77].

Recent Advancements in the Electrode Material

Electrochemical cells (SCs) consist of two electrodes: an anode and a cathode. The choice of electrode material is crucial for determining optimal voltage ranges for devices. [78] Several methodologies are employed to augment voltage ranges in SCs, with carbon-based electrodes being advantageous due to their chemical stability, C_{sp}, and high P_d (power density). [79] Graphene has emerged as a highly promising material for electrochemical applications, due to its exceptional structural integrity and outstanding electrical conductivity. However, its low E_d restricts its potential for many applications. [80, 81] Combining conductive polymer-based pseudocapacitors (TMO) with graphene has emerged as a viable approach to address various challenges. [82-84] Nanostructured TMOs have garnered considerable scientific attention as promising electrode materials for SC applications, owing to their superior structural attributes and favorable electrical properties. Their elevated Csp is largely attributed to the enhanced redox activity of TMOs compared to conventional carbon-based electrodes. Ongoing research into the capacitive behavior of TMOs continues to advance, driven by their intrinsically high charge-storage capabilities.

Numerous studies have provided detailed insights into the deployment of TMO composites in pseudocapacitor systems. Notably, Wei et al. synthesized hetero structured Co_3O_4 nanomaterials—comprising interconnected nanowires and nanosheets—that achieved a remarkable specific capacitance of 2053 $F \cdot g^{-1}$ along with outstanding cycling stability. Liu and his team used the direct cyclic voltammetry technique and in situ generation to fabricate CuO nanorods, which exhibited remarkable capacitance performance with a retention rate of 96.45% even after 4000 cycles. [85-87] Recent studies have demonstrated that bimetallic oxides may offer superior capacitance performance compared to their single-metal counterparts. This enhancement is attributed to synergistic interactions between the two metal components, which can improve redox activity, electronic conductivity, and structural stability in supercapacitor electrode materials. Farooq and his co-workers synthesized $Sr_2Ni_2O_5$/rGO (reduced graphene oxide) composites using a solvothermal method, resulting in enhanced electrochemical properties. [88,89]

Fabrication of Electrodes

The efficiency of solar cells (SCs) is strongly governed by the choice and optimization of electrode materials. These devices generally comprise of a thin film of photoactive compounds interfaced with a current collector, where the thickness of the active layer plays a critical role in determining overall performance parameters such as light absorption, charge transport, and energy conversion efficiency. The fabrication process for an electrode involves a conductive surface that facilitates electron transfer within or out of an electrochemical cell. Farooq and co-researchers engineered an electrode incorporating TMO /rGO composites for application in supercapacitor (SC) systems. The process involved three stages: etching nickel foam samples with a solution of H_2SO_4 and distilled water, washing and drying the etched pieces, and formulating a composite with PVDF and DMF. The solution was agitated at 150 rpm for 30 minutes, followed by placement on a heated surface to ensure thorough homogenization. A slurry was subsequently formulated using a 1:1:8 weight ratio of binding solution, activated carbon, and active material. The mass of the active material utilized in the mixture was precisely 0.008 g. Different TMO electrode materials were used, including $MgCo_2O_4$, IrO_2, NiO, $NiCo_2O_4$, MnO_2-rGO, and Co_3O_4/CNF. The prepared solution was subjected to thermal treatment on a heated plate for 8 hours prior to its application onto the nickel foam (NF) substrate. Following deposition, the sample was oven-dried to facilitate complete removal of residual moisture. The coated nanofibers were then used to further characterize the electrodes in SCs [90-96].

Transition Metal Oxide Composites: Pseudo-capacitive oxides of transition metals [97] are widely explored because of their high theoretical capacitances, low cost, and reversible faradaic redox reactions, leading to higher C_{sps} compared to carbonaceous materials based on an electrical double-layer charge storage mechanism. Despite their commendable electrochemical activity, transition metal oxides (TMOs) are inherently limited by poor electrical and ionic conductivity, leading to reduced rate capability and practical capacitance values that fall short of their theoretical potential. To address these constraints, extensive efforts have been made to engineer their nanostructures. Approaches include anchoring TMOs onto conductive substrates, doping with other metals to enhance both conductivity and redox behavior, and fabricating composite oxides featuring multiple oxidation states. Among these, coupling multi-valent oxide composites with conductive matrices has emerged as a particularly effective strategy for boosting electrical conductivity and, consequently, improving overall energy density.

Using MnOx-based composites as a representative system, the pseudocapacitance of MnO_2 arises from rapid, reversible surface redox transitions between Mn^{3+} and Mn^{4+} ions. The extremities of polarization potential are governed by irreversible redox pathways—specifically, the oxidation of Mn^{4+} to Mn^{7+} and its reduction to Mn^{2+}. Significant advancements in MnOx composite architectures have been reported. For instance, Qiu et al. developed a hybrid three-dimensional Au/MnOx nanocone array electrode via imprinting and soft-printing techniques, as illustrated in Figure 4. This unique array demonstrates multiple advantages, such as abundant active sites, a large electrode surface area, and facile electrolyte diffusion. Notably, the Au/MnOx composite encompasses complex phases including Mn_3O_4 and MnO–O–H, which facilitate proton transport in manganese oxide through hopping mechanisms between H_2O and OH sites, thereby enhancing ionic conductivity. As a result, the hybrid Au/MnOx electrode exhibited outstanding electrochemical performance, achieving a specific capacitance of 840.3 $F \cdot g^{-1}$ at a current density of 2 $A \cdot g^{-1}$—surpassing the capacitance values of many previously reported supercapacitor systems.

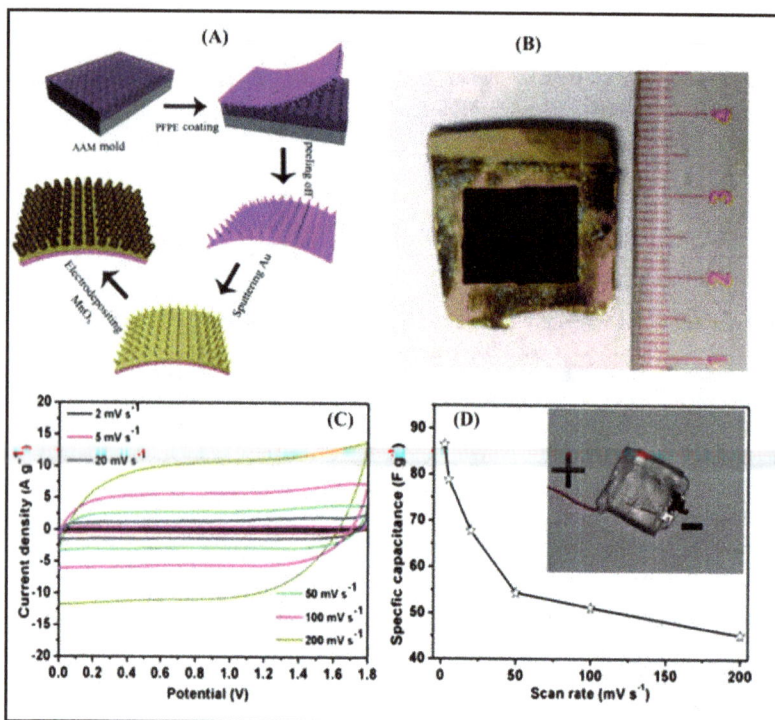

Figure 4. Fabrication of the 3D metal/oxide nanocone array electrode (a) Schematic diagram of the fabrication process of the 3D flexible nano cone array electrode, (b) Optical image of an as-prepared 3D Au/MnO_x nanocone electrode, (c) Performance testing of a full SC consisting of the 3D Au/MnO_x nanocone array positive electrode and a CCG negative electrode, (d) CV curves of the full SC at various scan rates in 1 M Na_2SO_4. Corresponding C_{sp} as a function of scan rates (Reproduced with permission from Ref. [97], Copyright 2020, Royal Society of Chemistry).

Ternary transition metal oxides ($M_xN_yO_z$) and their Composites: As previously discussed, although TMOs exhibit notable electrochemical activity, their application is often constrained by intrinsically low electrical conductivity and poor rate capability. To overcome these limitations, bimetallic oxides of the general form $M_xN_yO_z$_where M and N represent transition metals such as Zn, Ni, Mn, Co, or Cu—have attracted considerable attention. These materials demonstrate enhanced specific capacitance (Csp), stemming from improved conductivity, a greater abundance of active redox sites, and superior mechanical and thermal stability compared to their monometallic counterparts. For instance, nickel cobaltite (NiCo$_2$O$_4$) exhibits electrical conductivity that is two orders of magnitude higher than that of nickel, cobalt, or iron oxides, contributing to its excellent electrochemical performance. Building on this premise, iron–cobalt oxide (FeCo$_2$O$_4$), with its expanded redox activity, is considered a particularly promising candidate for high-performance energy storage applications.

Sethi M. et al. reported the synthesis of nickel ferrite (NiFe$_2$O$_4$) nanoparticles via a facile solvothermal method, resulting in a morphology characterized by high surface area. Electrochemical evaluation using a three-electrode configuration revealed a specific capacitance of 478 F·g^{-1} based on cyclic voltammetry (CV) at a scan rate of 5 mV·s^{-1}, and 368 F·g^{-1} at a current density of 1 A·g^{-1}. The electrode material also demonstrated a high energy density of 10.4 Wh·kg^{-1} and a power density of 225.8 W·kg^{-1} under identical current conditions. Long-term cycling performance was assessed over 10,000 charge–discharge cycles at 8 A·g^{-1}, with 88% of the initial capacitance retained—indicating robust electrochemical stability. The enhanced performance is attributed to the nanoparticles' nanoscale architecture featuring high porosity and surface area, which effectively facilitates electrolyte ion insertion/de-insertion while maintaining mechanical integrity during repeated cycling. Figure 5 gives the capacitance results of NiFe$_2$O$_4$ in three electrode and two electrode [98].

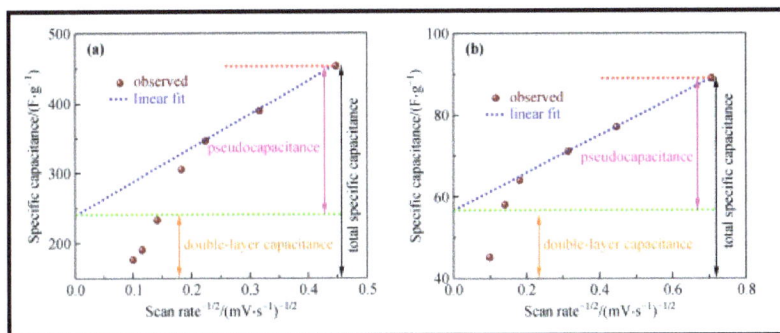

Figure 5. Graphical determination of capacitance contribution of NF in (a) the three-electrode system and (b) the fabricated supercapacitor device (Reproduced with permission from Ref. [98], Copyright 2020, Springer).

Zhu X. et al. developed a three-dimensional, hierarchical, self-supported hybrid electrode composed of NiO/Co$_3$O$_4$@C/CoS$_2$, wherein NiO/Co$_3$O$_4$ nanosheets were grown in situ on a nickel foam substrate and integrated with CoS$_2$ nanospheres via a conductive carbon interlayer. The synergistic effects arising from the porous architecture of NiO/Co$_3$O$_4$@C nanosheets and CoS$_2$ nanospheres significantly enhanced the electrochemical performance of the hybrid electrode. More recently, Chu et al. synthesized phosphorus-doped NiCo$_2$O$_4$ (P–NCO) with abundant oxygen vacancies through a hydrothermal process, followed by phosphatization in a tube furnace—resulting in improved ion transport and redox activity. The novel structure of NiCo$_2$O$_4$ (P–NCO) with greatly improved electrical conductivity shows an ultrahigh C_{sp} of 2747.8 Fg^{-1} at 1 Ag^{-1} and significant rate performance [99].

Conclusion

This chapter presented a comprehensive overview of recent research involving transition metal-based materials—particularly transition metal oxides and their composites—for high-performance SC applications. Owing to their high theoretical capacitance, cost-effectiveness, and strong electrochemical activity, transition metal compounds have emerged as compelling alternatives to conventional carbon-based materials such as activated carbon, biochar, and graphene. Nonetheless, the practical deployment of transition metal oxides is often hindered by their intrinsically low electrical conductivity. To overcome this limitation, researchers have explored the design and synthesis of binary metal oxides and composite systems, incorporating a secondary transition metal or other conductive components to enhance electrical performance. To extend to wide-scale commercial applications for energy storage, many challenges must be overcome when transition metal oxides and their compounds are synthesised and tested. In addition, the issue of environmental friendliness should also be addressed.

References

[1] C.N. Chervin, B.J. Clapsaddle, H.W. Chiu, A.E. Gash, J.H. Satcher, S. M. Kauzlarich, Chem. Mater. 18 (2006) 4865-4874. https://doi.org/10.1021/cm061258c

[2] P. Kanha, P. Saengkwamsawang, Inorg. Nano-Metal Chem. 47 (2017) 1129-1133. https://doi.org/10.1080/24701556.2017.1284100

[3] M. Saleem, F. Ahmad, M. Fatima, A. Shahzad, M.S. Javed, S. Atiq, M.A. Khan, M. Danish, O. Munir, S.M.B. Arif, U. Faryad, M.J. Shabbir, D. Khan, J. Energy Storage 76 (2024) 109822. https://doi.org/10.1016/j.est.2023.109822

[4] A.J. Ahamed, P. Vijaya KumarJ. Chem. Pharm. Res. 8 (2016) 624-628.

[5] Q. Geng, X. Su, F. Dong, Z. Wang, IOP Conf. Ser. Earth Environ. Sci. 706 (2021). https://doi.org/10.1088/1755-1315/706/1/012043

[6] F. Gaspar, C.D. Nunes, Catalysts 10 (2020). https://doi.org/10.3390/catal10020265

[7] T. Ghoshal, S. Biswas, M. Paul, S.K. De, J. Nanosci. Nanotechnol. 9 (2009) 5973-5980. https://doi.org/10.1166/jnn.2009.1290

[8] Y. Hu, H.J. Chen, J. Nanopart. Res. 10 (2008) 401-407. https://doi.org/10.1007/s11051-007-9264-0

[9] C.V. Krishnan, J. Chen, C. Burger, B. Chu, J. Phys. Chem. B 110 (2006) 20182-20188. https://doi.org/10.1021/jp063156f

[10] X. Li, S. Jiang, J. Li, K. Li, J. Li, J. Appl. Polym. Sci. 139 (2022). https://doi.org/10.1002/app.51849

[11] M. Vafaee, M.S. Ghamsari, Mater. Lett. 61 (2007) 3265-3268. https://doi.org/10.1016/j.matlet.2006.11.089

[12] R.S. Yadav, P. Mishra, A.C. Pandey, Ultrason. Sonochem. 15 (2008) 863-868. https://doi.org/10.1016/j.ultsonch.2007.11.003

[13] T. Yunusi, C. Yang, W. Cai, F. Xiao, J. Wang, X. Su, Ceram. Int. 39 (2013) 3435-3439. https://doi.org/10.1016/j.ceramint.2012.09.096

[14] M. Danish, M. Islam, F. Ahmad, M. Madni, M. Jahangeer, J. Phys. Chem. Solids 185 (2024) 111783. https://doi.org/10.1016/j.jpcs.2023.111783

[15] S. Ghorban Hosseini, Z. Khodadadipoor, Indian J. Chem. 57 (2018) 449-453.

[16] J. Huo, Y. Xue, Y. Liu, Y. Ren, G. Yue, J. Electroanal. Chem. 857 (2020) 113751. https://doi.org/10.1016/j.jelechem.2019.113751

[17] M. Karthi keyan, P. Vijaya Kumar, A. Jafar Ahamed, A. Ravikumar, J. Adv. Appl. Sci. Res. 2 (2021) 1-9. https://doi.org/10.46947/joaasr222020101

[18] Z. Wang, G.M. Kale, M. Ghadiri, J. Am. Ceram. Soc. 95 (2012) 3124-3129. https://doi.org/10.1111/j.1551-2916.2012.05366.x

[19] A. Wang, K. Sun, J. Li, W. Xu, J. Jiang, Mater. Chem. Phys. 231 (2019) 311-321. https://doi.org/10.1016/j.matchemphys.2019.04.046

[20] M. Zheng, X. Xiao, L. Li, P. Gu, X. Dai, H. Tang, H. Pang, Sci. China Mater. 61 (2) (2018) 185-209. https://doi.org/10.1007/s40843-017-9095-4

[21] J. Wang, Q. Zhong, Y. Xiong, D. Cheng, Y. Zeng, Y. Bu, Appl. Surf. Sci. 483 (2019) 1158-1165. https://doi.org/10.1016/j.apsusc.2019.03.340

[22] M.D. Angelin, S. Rajkumar, J.P. Merlin, A.R. Xavier, M. Franklin, A. T. Ravichandran, Ionics 26 (11) (2020) 5757-5772. https://doi.org/10.1007/s11581-020-03681-8

[23] M. Chaudhary, M. Singh, A. Kumar, Y.K. Gautam, A.K. Malik, Y. Kumar, B. P. Singh, Ceram. Int. 47 (2) (2021) 2094-2106. https://doi.org/10.1016/j.ceramint.2020.09.042

[24] D.P. Dubal, G.S. Gund, C.D. Lokhande, R. Holze, Mater. Res. Bull. 48 (2) (2013) 923-928. https://doi.org/10.1016/j.materresbull.2012.11.081

[25] Z. Song, W. Liu, N. Sun, W. Wei, Z. Zhang, H. Liu, Z. Zhao, Solid State Commun. 287 (2019) 27-30. https://doi.org/10.1016/j.ssc.2018.10.007

[26] Q. Wang, Y. Zhang, J. Xiao, H. Jiang, T. Hu, C. Meng, J. Alloys Compd. 782 (2019) 1103-1113. https://doi.org/10.1016/j.jallcom.2018.12.235

[27] P.S. Kumar, H.H. Kyaw, M.T.Z. Myint, L. Al-Haj, A.A.H. Al-Muhtaseb, M. Al-Abri, V.K. Ponnusamy, Int. J. Energy Res. 44 (13) (2020) 10682-10694. https://doi.org/10.1002/er.5712

[28] A. Qayyum, M. Okash, F. Ahmad, M. Ahmed, S.M. Ramay, S. Atiq, Solid State Ionics 395 (2023) 116227. https://doi.org/10.1016/j.ssi.2023.116227

[29] S. Aslam, S.M. Ramay, A. Mahmood, G.M. Mustafa, S. Zawar, S. Atiq, J. Sol-Gel Sci. Technol. 105 (2023) 360-369. https://doi.org/10.1007/s10971-022-06008-3

[30] H. Jin, J. Li, Y. Yuan, J. Wang, J. Lu, S. Wang, Adv. Energy Mater. 8 (23) (2018) 1801007. https://doi.org/10.1002/aenm.201801007

[31] Z.S. Iro, C. Subramani, S. Dash, Int. J. Electrochem. Sci. 11 (2016) 10628-10643. https://doi.org/10.20964/2016.12.50

[32] C.D. Lokhande, D.P. Dubal, O.-S. Joo, Curr. Appl. Phys. 11 (2011) 255e270. https://doi.org/10.1016/j.cap.2010.12.001

[33] M. Inagaki, H. Konno, O. Tanaike, J. Power Sources 195 (2010) 7880e7903. https://doi.org/10.1016/j.jpowsour.2010.06.036

[34] Nithiya S. George, Lolly Maria Jose and Arun Aravind, Chapter metrics Overview: Review on Transition metal oxides and their composites for energy storage application, 22 November 2022. https://doi.org/10.5772/intechopen.108781

[35] J.-K. Lee, H.M. Pathan, K.-D. Jung, O.-S. Joo, J. Power Sources 159 (2006) 1527e1531. https://doi.org/10.1016/j.jpowsour.2005.11.063

[36] C. Niu, E.K. Sichel, R.R. Hoch, D.D. Moy, H. Tennent, Appl. Phys. Lett. 70 (1997) 1480e1482. https://doi.org/10.1063/1.118568

[37] W.C. Fang, O. Chyan, C.L. Sun, C.T. Wu, C.P. Chen, K.H. Chen, L.C. Chen, J. H. Huang, Electrochem. Commun. 9 (2007) 239e244. https://doi.org/10.1016/j.elecom.2006.09.001

[54, 38] C.-C. Hu, K.-H. Chang, C.-C. Wang, Electrochim. Acta 52 (2007) 4411e4418. https://doi.org/10.1016/j.electacta.2006.12.022

[39] S. Yan, H. Wang, P. Qu, Y. Zhang, Z. Xiao, Synth. Met. 159 (2009) 158e161. https://doi.org/10.1016/j.synthmet.2008.07.024

[40] X. Wang, W.B. Yue, M.S. He, M.H. Liu, J. Zhang, Z.F. Liu, Chem. Mater. 16 (2004) 799e805. https://doi.org/10.1021/cm035070u

[41] L. Li, K.H. Seng, H. Liu, I.P. Nevirkovets, Z. Guo, Electrochim. Acta 87 (2013) 801e808. https://doi.org/10.1016/j.electacta.2012.08.127

[42] J. Gao, M.A. Lowe, H.D. Abruna, Chem. Mater. 23 (2011) 3223e3227. https://doi.org/10.1021/cm201039w

[43] F. Naamoune, B. Messaoudi, A. Kahoul, N. Cherchour, A. Pailleret, H. Takenouti, Ionics 18 (2012) 365e370. https://doi.org/10.1007/s11581-011-0621-8

[44] G. Laugel, J. Arichi, H. Guerba, M. Molire, A. Kiennemann, F. Garin, B. Louis, Catal. Lett. 125 (2008) 14e21. https://doi.org/10.1007/s10562-008-9523-4

[45] S.W. Donne, A.F. Hollenkamp, B.C. Jones, J. Power Sources 195 (2010) 367e 373. https://doi.org/10.1016/j.jpowsour.2009.06.103

[46] E.C. Rios, A.V. Rosario, R.M.Q. Mello, L. Micaroni, J. Power Sources 163 (2007) 1137e1142. https://doi.org/10.1016/j.jpowsour.2006.09.056

[47] G.A. Kriegsmann, J. Appl. Phys. 71 (1992) 1960e1966. https://doi.org/10.1063/1.351191

[48] M. Ghaemi, F. Ataherian, A. Zolfaghari, S.M. Jafari, Electrochim. Acta 53 (2008) 4607e4614. https://doi.org/10.1016/j.electacta.2007.12.040

[49] Z. Fan, J. Chen, B. Zhang, B. Liu, X. Zhong, Y. Kuang, Diamond Relat. Mater. 17 (2008) 1943e1948. https://doi.org/10.1016/j.diamond.2008.04.015

[50] G.A. Tompsett, W.C. Conner, K.S. Yngvesson, ChemPhysChem 7 (2006) 296e 319. https://doi.org/10.1002/cphc.200500449

[51] L. Chen, D. Zhu, Solid State Sci. 27 (2014) 69e72. https://doi.org/10.1016/j.solidstatesciences.2013.11.001

[52] N. Chen, K. Wang, X. Zhang, X. Chang, L. Kang, Z.-H. Liu, Colloids Surf. A 387 (2011) 10e16. https://doi.org/10.1016/j.colsurfa.2011.07.003

[53] S. Chou, F. Cheng, J. Chen, J. Power Sources 162 (2006) 727e734. https://doi.org/10.1016/j.jpowsour.2006.06.033

[54] D.P. Dubal, D.S. Dhawale, T.P. Gujar, C.D. Lokhande, Appl. Surf. Sci. 257 (2011) 3378e3382. https://doi.org/10.1016/j.apsusc.2010.11.028

[55] N. Behm, D. Brokaw, C. Overson, D. Peloquin, J. Poler, J. Mater. Sci. 48 (2013) 1711e1716. https://doi.org/10.1007/s10853-012-6929-6

[56] P. Justin, S.K. Meher, G.R. Rao, J. Phys. Chem. C 114 (2010) 5203e5210. https://doi.org/10.1021/jp9097155

[57] Wei Yu, Xinbing Jiang, Shujiang Ding, Ben Q. Li, J. Power Sources,256, June 2014, 440 - 448. https://doi.org/10.1016/j.jpowsour.2013.12.110

[58] Shankar G. Randive, H. M. Pathan and Balkrishna J. Lokhande, ES Energy & Environment, April 2023, 20, 877

[59] W. Li, T. Li, X. Ma, Y. Li, L. An, Z. Zhang, RSC Adv. 6 (15) (2016) 12491-12496. https://doi.org/10.1039/C5RA25550C

[60] T.W. Chen, S. Chinnapaiyan, S.M. Chen, M.A. Ali, M.S. Elshikh, A.H. Mahmoud, Sonochem. 63 (2020) 104903. https://doi.org/10.1016/j.ultsonch.2019.104903

[61] T.W. Chen, U. Rajaji, S.M. Chen, R.J. Ramalingam, X. Liu, 58 (2019) 104595. https://doi.org/10.1016/j.ultsonch.2019.05.012

[62] T.W. Chen, A. Sivasamy Vasantha, S.M. Chen, D.A. Al Farraj, M. Soliman Elshikh, R.M. Alkufeidy, M.M. Al Khulaifi, 59 (2019) 1-8. https://doi.org/10.1016/j.ultsonch.2019.104718

[63] F.M. Courtel, Y. Abu-Lebdeh, I.J. Davidson, Electrochim. Acta 71 (2012) 123-127. https://doi.org/10.1016/j.electacta.2012.03.108

[64] S.M. Ibrahim, S.A. Halim, J. Mol. Liq. 339 (2021) 116652. https://doi.org/10.1016/j.molliq.2021.116652

[65] H. Qiu, T. Du, J. Wu, Y. Wang, J. Liu, S. Ye, S. Liu, Dalton Trans. 47 (2018) 6934-6941. https://doi.org/10.1039/C8DT00893K

[66] M. Ranjbar-Azad, M. Behpour, J. Mater. Sci. Mater. Electron. 32 (2021) 18043-18056. https://doi.org/10.1007/s10854-021-06346-y

[67] E. Tamilalagan, M. Akilarasan, S.M. Chen, T.W. Chen, Y.C. Huang, Q. Hao, W. Lei, Ultrason. Sonochem. 67 (2020) 105164. https://doi.org/10.1016/j.ultsonch.2020.105164

[68] M. Vangari, T. Pryor, L. Jiang,J. Energy Eng. 139 (2013) 72-79. https://doi.org/10.1061/(ASCE)EY.1943-7897.0000102

[69] M. Setoodehkhah, S. Momeni, 18th Iranian Chemistry Congress- Semnan.

[70] Yong X. Gan , Ahalapitiya H. Jayatissa,Zhen Yu, Xi Chen , and Mingheng L, Journal of Nanomaterial, January 202

[71] L. Peng, Z. Fang, Y. Zhu, C. Yan, G. Yu, Adv. Energy Mater. 8 (2018) 1-19. https://doi.org/10.1002/aenm.201702179

[72] P.T. Shibeshi, D. Parajuli, N. Murali, Chem. Phys. 561 (2022) 111617. https://doi.org/10.1016/j.chemphys.2022.111617

[73] S. Tajik, H. Beitollahi, Z. Dourandish, P. Mohammadzadeh Jahani, I. Sheikhshoaie, M.B. Askari, P. Salarizadeh, F.G. Nejad, D. Kim, S.Y. Kim, R. S. Varma, M. Shokouhimehr, Electroanalysis 34 (2022) 1065-1091. https://doi.org/10.1002/elan.202100393

[74] D. Wang, C. Duan, H. He, Z. Wang, R. Zheng, H. Sun, Y. Liu, C. Liu, J. Colloid Interface Sci. 646 (2023) 89-97. https://doi.org/10.1016/j.jcis.2023.05.043

[75] D. Yassen, F. Othman, A. Abdul Hamead, Eng. Technol. J. 40 (2022) 862-868. https://doi.org/10.30684/etj.v40i6.2104

[76] X. Yin, J. Liu, J. Ma, C. Zhang, P. Chen, M. Que, Y. Yang, W. Que, C. Niu, J. Shao, J. Power Sources 329 (2016) 398-405. https://doi.org/10.1016/j.jpowsour.2016.08.102

[77] T.R. Bastami, M.H. Entezari,. Sonochem. 19 (2012) 830-840. https://doi.org/10.1016/j.ultsonch.2011.11.019

[78] J.H. Lee, G. Yang, C.-H. Kim, R.L. Mahajan, S.-Y. Lee, S.-J. Park, Energy Environ. Sci. 15 (2022) 2233-2258. https://doi.org/10.1039/D1EE03567C

[79] Q. Ma, M. Liu, F. Cui, J. Zhang, T. Cui, Carbon 204 (2023) 336-345. https://doi.org/10.1016/j.carbon.2022.12.066

[80] S. Lin, J. Tang, K. Zhang, T.S. Suzuki, Q. Wei, M. Mukaida, Y. Mukaida, H. Zhang, X. Mamiya, L.C. Qin Yu, J. Power Sources 482 (2021). https://doi.org/10.1016/j.jpowsour.2020.228995

[81] T. Doi, Y. Shimizu, M. Hashinokuchi, M. Inaba, J. Electrochem. Soc. 163 (2016) A2211. https://doi.org/10.1149/2.0331610jes

[82] Y. Chen, C. Kang, L. Ma, L. Fu, G. Li, Q. Hu, Q. Liu, Chem. Eng. J. 417 (2021) 129243. https://doi.org/10.1016/j.cej.2021.129243

[83] C. Guan, A. Sumboja, H. Wu, W. Ren, X. Liu, H. Zhang, Z. Liu, C. Cheng, S. J. Pennycook, J. Wang, Adv. Mater. 29 (44) (2017). https://doi.org/10.1002/adma.201704117

[84] L. Shahriary, A.A. Athawale, Int. J. Renew. Energy Environ. Eng. 2 (1) (2014) 58-63. https://doi.org/10.1155/2014/903872

[85] Y. Ji, Y. Deng, F. Chen, Z. Wang, Y. Lin, Z. Guan, Carbon 156 (2020) 359-369. https://doi.org/10.1016/j.carbon.2019.09.064

[86] Y. Lu, L. Li, D. Chen, G. Shen, J. Mater. Chem. A 5 (47) (2017) 24981-24988. https://doi.org/10.1039/C7TA06437C

[87] Y. Ouyang, X. Xia, H. Ye, L. Wang, X. Jiao, W. Lei, Q. Hao, ACS Appl. Mater. Interfaces 10 (4) (2018) 3549-3561. https://doi.org/10.1021/acsami.7b16021

[88] G. Wei, L. Yan, H. Huang, F. Yan, X. Liang, S. Xu, Z. Lan, W. Zhou, J. Guo, Appl. Surf. Sci. 538 (2021) 147932. https://doi.org/10.1016/j.apsusc.2020.147932

[89] Y. Liu, X. Cao, D. Jiang, D. Jia, J. Liu, J. Mater. Chem. A 6 (22) (2018) 10474-10483. https://doi.org/10.1039/C8TA00945G

[90] V. Modafferi, S. Santangelo, M. Fiore, E. Fazio, C. Triolo, S. Patan' e, R. Ruffo, M. G. Musolino, Transition metal oxides on reduced graphene oxide nanocomposites: evaluation of physicochemical properties, J. Nanomater. 2019 (2019). https://doi.org/10.1155/2019/1703218

[91] A.A. Yadav, Influence of electrode mass-loading on the properties of spray deposited Mn3O4 thin films for electrochemical supercapacitors, Thin Solid Films 608 (2016) 88-96. https://doi.org/10.1016/j.tsf.2016.04.023

[92] G. Yu, L. Hu, M. Vosgueritchian, H. Wang, X. Xie, J.R. McDonough, X. Cui, Y. Cui, Z. Bao, Solution-processed graphene/MnO2 nanostructured textiles for high- performance electrochemical capacitors, Nano Lett. 11 (2011) 2905-2911. https://doi.org/10.1021/nl2013828

[93] B.K. Kim, S. Sy, A. Yu, J. Zhang, Electrochemical supercapacitors for energy storage and conversion, in: Handbook of Clean Energy Systems, John Wiley & Sons, Ltd., 2015, pp. 1-25. https://doi.org/10.1002/9781118991978.hces112

[96] Z. Cao, Y. Zhang, Y. Cui, J. Gu, Z. Du, Y. Shi, K. Shen, H. Chen, B. Li, S. Yang, Harnessing the unique features of 2D materials toward dendrite-free metal anodes, Energy Environ. Mater. 5 (2022) 45-67. https://doi.org/10.1002/eem2.12165

[97] Mingjin Cui and Xiangkang Meng Overview of transition metal-based composite materials for supercapacitor electrodes Nanoscale Adv., 2020, 2,5516. https://doi.org/10.1039/D0NA00573H

[98] Meenaketan Sethi, U. Sandhya Shenoy, Selvakumar Muthu, and D. Krishna Bhat, Front. Mater. Sci., February 2020

[99] Xingxing Zhu, Mengyao Sun, Rui Zhao, Yingqi Li, Bo Zhang, Yingli Zhang, Xingyou Lang, Yongfu Zhu and Qing Jiang, J. Nanoscale Advances, 7, May 2020

Electrocatalysts and Advanced Materials for Sustainable Energy Storage   Materials Research Forum LLC
Materials Research Foundations 182 (2025) 118-134   https://doi.org/10.21741/9781644903797-9

Chapter 9

Carbonaceous Materials in Electrocatalysis: Innovations and Applications towards Sustainable Energy Storage

DICKSON D. Babu[1,a*], PRAVEEN Naik[2,b]

[1]Department of Chemistry, St. Thomas College, Kozhencherry, 689641, Kerala, India

[2]Department of Chemistry, Nitte Meenakshi Institute of Technology, Nitte (Deemed to be University), Bengaluru Campus, Bengaluru 560064, India

[a]dicksondbabu@stthomascollege.info, [b]praveennaik018@gmail.com

Abstract

The development of efficient and sustainable electrocatalysts is crucial for advancing energy storage and conversion technologies. While noble metals (e.g., Pt, Pd, Ir, Ru) and transition metal-based compounds (e.g., FeO, MoS_2) have been widely explored, their high cost, limited durability, and environmental concerns have driven the search for alternative materials. Carbonaceous materials, especially metal-free and heteroatom-doped structures, have recently appeared as very efficient and cost-effective electrocatalysts with remarkable performance in fuel cells, metal–air batteries, and catalytic water splitting. This chapter offers a review of the latest breakthroughs in carbon-based electrocatalysis, focusing particularly on two-dimensional (2D) materials. This includes exploration into the fine engineering of active sites via thermal condensation, roles of heteroatom doping and defect modulation, and understanding structure-activity relationships which regulate catalytic efficiency. The mechanistic discussion concerning 2D structural growth also encompasses the impact upon electrocatalytic functionality. Finally, the chapter summarizes some of the major challenges and future opportunities in incorporating carbonaceous materials into next-generation energy storage and conversion systems toward sustainable and scalable applications.

Keywords

Carbonaceous Materials, Electrocatalysts, Energy Storage, 2D-Materials

Contents

Introduction

The world's energy map is undergoing a dramatic and necessary shift to lower carbon emissions and combat climate change. Solar and wind power have increased in prominence, but effective energy storage and conversion technologies are still necessary for their large-scale deployment. Electrocatalysis has a central function in enabling important reactions in energy storage and conversion devices, such as water splitting, fuel cells, and CO_2 reduction [1,2]. Through improving the kinetics and efficiency of the reactions, electrocatalysts lead to the innovation of sustainable and high-performance energy technologies.

On the quest for cost-effective, eco-friendly alternatives to noble metal-based electrocatalysts, carbonaceous materials have become the most promising candidates. Their singular attributes, such as high electrical conductivity, tunable surface chemistry, and structural diversity, make them the preferred candidates for many electrochemical applications. Graphene, carbon nanotubes (CNTs), porous carbons, and heteroatom-doped carbon frameworks have been found to possess superior catalytic activity in fuel cells, metal-air batteries, and water-splitting reactions. Their electrocatalytic function can also be engineered further through doping, creation of defects, and nano structuring.

Recent developments in material science have made it possible to develop new carbon-based electrocatalysts with improved efficiency and stability. New synthesis approaches, like defect engineering, heteroatom doping, and hybridization with metal and metal oxide nanoparticles, have offered nanoscale engineering of electronic properties and surface reactivity with high accuracy. Moreover, the coupling of machine learning and computational modelling has greatly facilitated the discovery and optimization of electrocatalyst structures, narrowing the gap between theoretical prediction and experimental realization.

This chapter offers a comprehensive investigation of the design principles, mechanistic understanding, and practical applications of carbon-based electrocatalysts. Some of the key issues pertain to the intrinsic electronic and structural properties of carbonaceous materials, their functioning in different electrochemical reactions, and recent trends in synthesis and functionalization methodology. Particular focus is laid on the part played by two-dimensional (2D) material and heteroatom doping for maximizing electrocatalytic efficiency. The industrial scalability of the materials is also addressed, and their cost viability and eco-friendliness are taken into consideration. Lastly, the chapter identifies present challenges and potential future research avenues, with a view to understanding comprehensively how carbonaceous materials can be leveraged for future energy storage and conversion technologies.

Fundamentals of Carbon-Based Electrocatalysis

The interest toward carbon-based electrocatalysts has grown with respect to their abundance, chemical tunability, environmentally friendly nature, and extraordinary physicochemical attributes. Unlike conventional noble-metal systems, carbonaceous materials can be changed at the atomic and molecular levels to control their catalytic properties through diverse electrochemical reactions in energy conversion, storage, etc. This section presents the structural and electronic properties of carbon materials and their mechanistic roles in electrocatalysis, with the crucial parameters that affect their efficiencies.

Structural and Electronic Properties of Carbonaceous Materials: Carbon-based materials possess diverse morphologies and hybridization states (sp, sp^2, sp^3), which confer upon them unique electronic and structural properties. Among these, sp^2-hybridized carbon structures, such as graphene, carbon nanotubes (CNTs), fullerenes, and graphitic carbon, are particularly important in electrocatalysis due to their excellent electrical conductivity coupled with delocalized π-electron systems and mechanical strength [3,4]. Graphene, for example, has a two-dimensional (2D) structure with a high theoretical surface area ($\sim$2630 m^2/g) and excellent electron mobility. On the other hand, carbon nanotubes act as one-dimensional (1D) conductive pathways with tunable wall thickness and diameters, thus serving as excellent charge transport materials in electrochemical systems. Activated carbon and mesoporous carbon provide high surface areas and rich pore architectures but are generally poor conductors unless hybridized with graphitic domains. A serious contender in this regard is heteroatom doping, in which atoms such as nitrogen (N), boron (B), sulphur (S), phosphorus (P), or fluorine (F) are introduced into the carbon matrix. These dopants induce charge neutrality in the carbon lattice, thereby altering the spin densities and inducing localized charge redistribution, which then leads to stronger interactions with reaction intermediates [5]. For nitrogen doping, different configurations are formed, such as pyridinic-N, pyrrolic-N, and graphitic-N, all of which play different roles in catalysis. Pyridinic-N can activate the ORR via stabilizing O_2 adsorption, while graphitic-N raises the electrical conductivity [6]. Also, defects and edges, such as vacancies, Stone–Wales defects, and topological distortions, become chemically active sites. Such defects could occur inherently or could be introduced during high-temperature treatments or chemical oxidation. While structural disorder is often considered detrimental to catalysis, it is found to often augment catalytic activity by creating unsaturated carbon atoms that become more reactive towards small molecules such as O_2, H_2O, or H^+ [7].

Electrocatalysts and Advanced Materials for Sustainable Energy Storage   Materials Research Forum LLC
Materials Research Foundations 182 (2025) 118-134   https://doi.org/10.21741/9781644903797-9

Mechanistic Insights into Carbon-Based Electrocatalysis: Understanding the catalytic mechanism is crucial for rationally designing carbon-based electrocatalysts. Broadly, these materials contribute to electrocatalysis in two ways:

(i) As metal-free active sites: Metal-Free Electrocatalysis in metal-free systems, electrocatalytic activity originates from doped heteroatoms, edge defects, and other local chemical environments within the carbon lattice. For instance, in the oxygen reduction reaction (ORR), a four-electron pathway is desirable for efficiency. Nitrogen-doped graphene has been shown to favour such a mechanism by enhancing O–O bond cleavage through electron-rich sites that stabilize key intermediates such as OOH^*, O^*, and OH^* [8]. Moreover, the density and nature of doped sites directly affect the reaction pathway. Pyridinic-N can attract O_2 molecules because of its lone pair electrons and can activate them through a decrease in the energy barrier for bond dissociation [9]. For the hydrogen evolution reaction (HER), one of the vital descriptors is the Gibbs free energy of hydrogen adsorption (ΔG_H^*). Therefore, the best electrocatalyst should have $\Delta G\_H^* \approx 0$. Doped carbon materials, mainly those functionalized with N or S, impart a unique flexible electronic adaptability that could modulate ΔGH^* to enhance competitiveness with conventional Pt-based catalysts [10].

(ii) Carbon as Support: Carbon is also the host of choice for catalytically reactive species, such as the transition metal nanoparticles or single-atom catalysts, due to its high surface and chemical stability, which would provide for finely even distribution and firm anchoring of metal species against agglomeration and leaching during catalytic operation [10-12]. Furthermore, this strong electronic interaction between the carbon support and the active metal species can arrange the d-band centre of metals in such a way as to improve their binding strength to intermediates and consequently the turnover frequencies[13].

Factors Influencing Electrocatalytic Activity: Carbon-based electrocatalysts exhibit performance characteristics that depend on a delicate balance of physicochemical properties, many of which are tuneable during both synthesis and post-synthesis treatment.

(i) Doping Chemistry: The identity and configuration of dopants play a role in modifying local charge densities and reactivity. For example, B-doping serves to add electron-deficient sites that operate as Lewis acid centres, while the opposite is true for N-doping because it situates electron-rich sites that favour redox reactions [9]. Dual-doping, such as N and S, could bring in synergistic effects whereby one dopant modifies the charge distribution while the other one adjusts the spin density in order to further promote the activity.

(ii) Structural Defects: The defined introduction of defects, including vacancies, pores, and disordered regions, generates active sites without hindering conductivity. Such defects could be produced thermally or chemically and are particularly beneficial for reactions of the adsorbate-desorbate type, such as ORR and CO_2 reduction [7].

(iii) Porosity and Surface Area: Mesopores (2–50 nm pore diameter) are favourable for the rapid mass transport of electrolytes and provide better access to active sites. High BET surface area guarantees that maximum reactive centres will be exposed to the electrolyte interface [14].

(iv) Electrical Conductivity: Conductivity controls the electron flow toward and from catalytic sites. Though amorphous carbon has poor conductivity, hybridization with graphitic domains or metallic nanoparticles can boost its performance. This is important, especially for high-rate applications like supercapacitors or water electrolysis [4].

(v) Wettability and Surface: Functionality: Among functional groups, hydroxyl, carboxyl, and carbonyl affect material interaction with aqueous electrolytes. Moderate hydrophilicity facilitates electrolyte accessibility and stabilization of the double layer to improve charge transfer kinetics [15]. However, excessive functionalization can impede conductivity and should be optimized carefully.

Types of Carbonaceous Electrocatalysts

Carbon-based materials have garnered significant attention as oxygen electrocatalysts due to their affordability, high availability, excellent electrical conductivity, and strong resistance to alkaline electrolytes. These properties contribute to enhanced oxygen reduction reaction (ORR) and oxygen evolution reaction (OER) efficiency. However, the inert nature of pristine carbon surfaces limits the adsorption and activation of reaction intermediates. To overcome this limitation, strategies such as chemical doping and the introduction of structural defects have been explored to improve the electrocatalytic performance of carbon nanomaterials.

Graphene and Graphene Oxide-Based Catalysts: Graphene and its derivatives offer a promising solution as conductive scaffolds for hybrid materials with electrocatalysts, owing to their unique structure and exceptional physical properties. These derivatives, obtained through various synthetic methods, exhibit diverse characteristics in lateral size, morphology, number of layers, and defect density. The production method plays a crucial role in determining their final properties. However, high costs and limited scalability restrict the use of techniques like mechanical exfoliation and chemical vapor deposition (CVD) to fundamental research and academic applications.

Among scalable approaches, the reduction of graphene oxide (GO) is widely adopted. GO is typically synthesized by oxidizing pristine graphite, followed by strong physical energy treatments such as ultrasonication or stirring in a liquid medium, enabling bulk production with high yield and reduced costs. However, GO has a high density of defects with disrupted sp^2 bonding networks. Graphene oxide (GO) is typically reduced via chemical, thermal, or electrochemical processes to regenerate its π-electron system and improve conductivity, resulting in reduced graphene oxide (rGO). Although some defects persist on the structure after reduction, these serve as functionalization or hybridization sites with electroactive materials, thus aiding the versatility of rGO-based composites.

Another benefit of rGO to the construction of composite materials using chemical redox routes is its ease in dispersion in various solvents. Such composites will find application directly or might be further reduced to gain rGO's conductive properties. In recent years, rGO has become a focus of attention for electrochemical energy conversion reactions. The literature survey shows that rGO has been largely exploited as an electron mediator in water-based conversion reactions: hydrogen evolution reaction (HER), oxygen evolution reaction (OER), and oxygen reduction reaction (ORR). While ORR research is stagnating, the study of water-splitting reactions is gaining momentum. Interest in rGO-based electrocatalysts for CO_2 reduction reaction (CO_2RR) and nitrogen reduction reaction (NRR) is still nascent but is witnessing rapid growth. The emerging trends undoubtedly point out the potential of rGO as a versatile material for electrochemical energy conversion toward sustainable and renewable energy technologies.

Electrocatalysts and Advanced Materials for Sustainable Energy Storage   Materials Research Forum LLC
Materials Research Foundations 182 (2025) 118-134   https://doi.org/10.21741/9781644903797-9

Carbon Nanotubes (CNTs) in Electrocatalysis

Carbon-based materials are considered promising electrocatalysts due to their structural versatility, high electrical conductivity, and chemical stability. Among these, carbon nanotubes (CNTs) cylindrical nanostructures composed of sp^2-hybridized carbon atoms arranged in a graphitic lattice have been widely studied for their unique electronic and electrocatalytic properties. CNTs are generally classified into single-walled (SWCNTs) and multi-walled (MWCNTs) types, with their edge configurations (armchair, zigzag, and chiral) significantly influencing their catalytic behaviour

Despite their exceptional mechanical strength and electrical conductivity, pristine CNTs exhibit low intrinsic catalytic activity owing to their chemically inert graphitic framework. To improve their electrocatalytic performance, various modification strategies have been employed, including defect engineering, heteroatom doping, and surface functionalization. Defect engineering introduces vacancies, edge distortions, and topological irregularities, leading to localized charge redistribution and the formation of active sites for electrochemical reactions. Heteroatom doping in which carbon atoms are replaced by elements such as nitrogen, boron, sulfur, or phosphorus modifies the electronic structure, thereby enhancing conductivity and catalytic activity.

Surface functionalization, often achieved by anchoring metal or non-metal species onto the CNT surface, promotes synergistic interactions that improve catalytic selectivity and stability. Additionally, CNTs are frequently hybridized with other catalytic materials, such as metal nanoparticles, metal oxides, or single-atom catalysts. These hybrid systems leverage the high surface area and conductivity of CNTs along with the catalytic properties of the incorporated materials. The resulting interfacial charge transfer enhances electron transport and reaction kinetics, while the tubular structure of CNTs provides a scaffold that regulates nucleation and prevents agglomeration of the active species. In some cases, CNTs can also confine nanoscale catalysts within their hollow cores, stabilizing active sites and improving catalytic efficiency through spatial confinement.

CNT-based electrocatalysts have been extensively investigated for key energy conversion reactions, including the oxygen reduction reaction (ORR) in fuel cells, the hydrogen evolution reaction (HER) and oxygen evolution reaction (OER) in water splitting, and the carbon dioxide reduction reaction (CO$_2$RR) for electrochemical CO_2 conversion. Recent advancements in synchrotron-based spectroscopic techniques have enabled in-depth analysis of the dynamic structural evolution of CNT-based electrocatalysts under operating conditions. High-resolution X-ray absorption spectroscopy (XAS) and in situ transmission electron microscopy (TEM) have proven especially useful in identifying real-time transformations in active sites and electronic structures, thereby deepening the understanding of structure–activity relationships and catalyst degradation mechanisms.

Despite these advancements, several challenges remain in optimizing CNT-based electrocatalysts for practical applications. These include achieving precise control over defect density and heteroatom distribution, enhancing long-term electrochemical stability, and developing scalable synthesis methods for uniform dispersion of catalytic species. Future research should focus on the integration of CNTs with other emerging materials, such as graphene and MXenes, to develop next-generation electrocatalysts. Additionally, computational modelling and machine learning techniques can play a key role in accelerating the design and optimization of CNT-based materials for sustainable energy applications.

Porous Carbon and Carbon Quantum Dots

Carbon quantum dots (CQDs) and porous carbon are considered promising materials for energy storage and electrocatalysis due to their high surface area, tunable porosity, and abundant surface functional groups. CQDs, as a type of zero-dimensional nanocarbon material, possess superior electrical conductivity, fast electron transfer, and rich active sites, and therefore are perfect candidates for supercapacitors and metal-ion batteries. Their surface functional groups allow for strong interactions with other active materials, promoting composite formation with metal oxides and conductive polymers to improve specific capacity, cycle stability, and rate performance. Moreover, heteroatom doping of CQDs efficiently tunes charge distribution and optimizes the electronic structure, enhancing considerably their electrocatalytic ability for oxygen reduction (ORR), oxygen evolution (OER), and hydrogen evolution reactions (HER), which are vital for their applications in fuel cells, metal–air batteries, and water electrolysis.

The recent breakthroughs in the synthesis methods have facilitated large-scale, low-cost production of CQDs with optimal properties. Techniques like NaOH-assisted processing of organic precursors and molecular fusion techniques have proved scalable synthesis without the use of toxic chemicals or high energy consumption. Such developments have promoted the use of CQDs in multifunctional energy storage and conversion systems. In addition, the incorporation of CQDs with porous carbon materials facilitates ion diffusion, charge transfer, and electrocatalytic efficiency, opening up next-generation energy solutions. Although notable advancements have been made, critical challenges still hinder the development of carbon quantum dots (CQDs), particularly in terms of achieving uniform control over their size, tailoring their surface functionalities, and ensuring long-term structural stability. Overcoming these limitations through innovative approaches in material design and synthesis will play a vital role in advancing the application of CQDs and porous carbon materials in energy-related technologies.

Doped and Functionalized Carbon Materials

Carbon based materials modified by heteroatom doping and surface functionalization have emerged as promising alternatives for conventional noble metal based electrocatalysts. These materials promise numerous advantages such as enhanced catalytic activity, higher durability, and lowered cost. The integration of heteroatoms, such as nitrogen (N), phosphorus (P), and sulfur (S), into the carbon lattice resulted in introducing electronic heterogeneity and novel active sites, which dramatically improve the performance in oxygen reduction reaction ORR, oxygen evolution reaction OER, and hydrogen evolution reaction HER. For example, N doping of graphene increases ORR activity by increasing electron density or inducing charge polarization at given sites. In the same way, S and P doping tune electronic structure and modulate their adsorption energies, which are important features of catalytic behaviour. These doped carbons are good replacements for noble metal catalysts as they are significantly stable and turn out to be less easily destroyed under corrosive conditions and operational use for a longer time period.

Functionalization techniques such as surface tailoring, metal coordination, or hierarchical nano structuring techniques would further help amplify catalytic efficiency. From the structural and catalytic point of view, metal-organic frameworks (MOFs) and their derivatives after carbonization serve as versatile scaffolds to construct porous carbon architectures with very precise control on functionality. Often used in MOF-derived carbons are the comprehensive transition metals: iron (Fe), cobalt (Co), or nickel (Ni). The high surface area and atomically dispersed active sites of such carbons contribute greatly to the significant electrocatalytic activity. Besides, other types of modified forms of graphene and carbon nanotubes incorporated with catalytic centres or decorated with nanoparticles have shown excellent performance in energy storage and conversion systems. Heteroatom doping combined with functionalization strategies would offer researchers an opportunity to design carbon-based electrocatalysts with fine-tuned properties for emerging energy technologies like fuel cells, metal-air batteries, and supercapacitors.

Electrocatalytic Applications of Carbonaceous Materials

Carbonaceous electrocatalysts are therefore central to energy conversion technologies owing to their variable surface chemistries, high electrical conductivity, and adaptable structural construct. These materials promise innovations in many electrochemical reactions, including the oxygen reduction (ORR), oxygen evolution (OER), hydrogen evolution (HER), and CO_2 reduction (CO_2RR) processes. With very good performance in these key reactions, they stand as important building blocks for the sustained, efficient energy system of the future.

Oxygen Reduction Reaction (ORR)

(i) Mechanistic Pathways and Active Sites: The oxygen reduction reaction (ORR) is a critical process in the operation of fuel cells and metal–air batteries; however, its practical application is frequently hindered by inherently slow reaction kinetics and the requirement for high overpotentials. Carbonaceous materials, particularly those doped with nitrogen, sulphur, or boron, can serve as effective metal-free ORR catalysts by promoting alternative reaction pathways.

Two primary ORR pathways are observed:

Four-electron ($4e^-$) pathway:

$$O_2 + 4H^+ + 4e^- \rightarrow 2H_2O \text{ (acidic) or } O_2 + 2H_2O + 4e^- \rightarrow 4OH^- \text{ (alkaline)}$$

Two-electron ($2e^-$) pathway:

$$O_2 + 2H^+ + 2e^- \rightarrow H_2O_2$$

The $4e^-$ pathway is more desirable due to its higher efficiency and avoids the formation of corrosive hydrogen peroxide. Pyridinic and graphitic nitrogen atoms embedded in the carbon lattice enhance the binding and activation of O_2, facilitating the $4e^-$ transfer route [6,8]. Other dopants like sulphur and boron alter the local electronic structure, creating spin density and charge delocalization that promote O_2 adsorption and activation. Defects, edge-plane sites, and curvature in carbon nanostructures also contribute significantly by providing high-energy sites that lower the activation barrier for O–O bond cleavage.

(ii) Applications in Fuel Cells

In proton exchange membrane fuel cells (PEMFCs) and alkaline fuel cells, carbon-based ORR catalysts provide a cost-effective and durable alternative to platinum. Nitrogen-doped graphene and carbon nanotubes have demonstrated high onset potentials, low peroxide yields, and excellent durability under acidic and alkaline conditions [6,9]. Additionally, carbon-supported metal–nitrogen–carbon (M–N–C) frameworks, where Fe or Co is coordinated to N-doped carbon matrices, offer superior activity and selectivity. These catalysts combine the high intrinsic activity of metal centres with the conductivity and corrosion resistance of carbon supports, making them highly effective in practical fuel cell devices. Notably, Huang et al. reported a versatile strategy for integrating multiple active sites into graphene through pyrolysis of hydrogen-bonded precursors, resulting in the Co-based hybrid catalyst [Figure 1a]. The resulting catalyst exhibited excellent trifunctional activity, enabling efficient performance in zinc–air batteries as well as overall water splitting. This work offers a general strategy for designing advanced multifunctional electrocatalysts for renewable energy applications, thereby simplifying the device design.[2]

Oxygen Evolution Reaction (OER)

OER is the anodic half-reaction in water electrolysis and is inherently sluggish due to its four-electron mechanism:

$2H_2O \rightarrow O_2 + 4H^+ + 4e^-$ (acidic) or $4OH^- \rightarrow O_2 + 2H_2O + 4e^-$ (alkaline)

Pure carbon materials typically suffer from low OER activity due to poor adsorption of oxygenated intermediates. However, when doped or hybridized with transition metal oxides or single-atom metal sites (e.g., Fe, Co, Ni) on N-doped carbon, their performance improves significantly. Carbon scaffolds enhance electron transport and offer high surface areas for active species dispersion. For instance, Han et. al. reported that nitrogen-doped hollow carbon nanocubes embedded with atomically dispersed cobalt and nickel dual sites (referred to as CoNi-SAs/NC) were synthesized through a pyrolysis process involving dopamine-coated metal-organic frameworks, which exhibited remarkable bifunctional activity for both oxygen reduction and evolution reactions. These properties enabled high-performance operation in practical rechargeable zinc–air batteries, delivering excellent energy efficiency, reduced overpotentials, and strong cycling stability, surpassing both conventional materials and leading noble-metal-based catalysts [16]. In alkaline water electrolyzers, these hybrid catalysts promote long-term operation without the corrosion issues associated with metal oxides alone. Here, the role of carbon is dual: it ensures mechanical stability and accelerates charge transfer, making the entire electrocatalytic system more efficient.

Hydrogen Evolution Reaction (HER)

HER is the cathodic half-reaction in water splitting:

$2H^+ + 2e^- \rightarrow H_2$ (acidic) or $2H_2O + 2e^- \rightarrow H_2 + 2OH^-$ (alkaline)

The challenge lies in developing catalysts with low overpotential and near-zero ΔG_H^* (free energy of H* adsorption). Metal-free carbon materials, such as N- and S-doped graphene or CNTs, can activate HER by altering the electronic structure of carbon and by the introduction of active sites with favourable ΔG_H^* values [10]. Hybrid systems, such as those based on MoS_2 nanosheets over N-doped carbon, have an enhanced activity due to the synergistic effect of MoS_2 edge sites along with the conductive carbon substrate. Carbon incorporation into transition metal phosphides (Chi-P, CoP) or carbides (Mo_2C) serves to enhance catalytic performance and structural durability. More recent developments also include single-atom catalysts (SACs), where isolated metal atoms are anchored on carbon supports. Such SACs, particularly those of Fe or Co in N-doped carbon matrices, have shown superior HER activity at low metal loadings, and hence, these are very attractive for large-scale hydrogen generation.

Electrocatalysts and Advanced Materials for Sustainable Energy Storage   Materials Research Forum LLC
Materials Research Foundations 182 (2025) 118-134   https://doi.org/10.21741/9781644903797-9

CO_2 Reduction Reaction (CO_2RR)

Electrocatalytic CO_2RR is a path for translating captured CO_2 into useful chemicals and fuels such as CO, HCOOH, CH_4, and C_2H_4. It has a considerable impact on closing the carbon cycle and subsequent mitigation efforts for greenhouse gas emissions. Carbon-based materials have been shown to be promisingly selective and stable for CO_2RR, especially when doped with N, B, or S. The nitrogen-doped graphene stabilizes key intermediates like COOH*, hence encouraging the formation of CO with a high Faradaic efficiency [17]. Functionalizing electrocatalysts based on carbon provides an interesting platform for the selective and efficient conversion of CO_2 to more reduced gaseous fuels such as carbon monoxide (CO), formic acid (HCOOH), methane (CH_4), ethylene (C_2H_4), and many carbonaceous products above that (C_2+). The CO_2RR process starts with the adsorption of CO_2 on the catalyst surface and then allows coupling between protons, followed by electron transfer steps, leading to a diverse range of reaction intermediates, such as *COOH, *CO, and *OCHO, which all depend on the catalyst structure and the operating conditions.

For example, the reaction to produce formate ($HCOO^-$), which is an important and wanted product, generally runs as follows:

$$CO_2 + H^+ + 2e^- \rightarrow HCOO^-$$

This reaction pathway involves the *OCHO intermediate and is favoured on catalyst surfaces that effectively stabilize oxygen-bound species. By contrast, the formation of carbon monoxide (CO) occurs via an alternative mechanism:

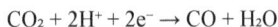

$$CO_2 + 2H^+ + 2e^- \rightarrow CO + H_2O$$

This mechanism involves the *COOH intermediate, which undergoes further reduction to yield CO. The selectivity between these two products (formate vs. CO) depends heavily on the nature and coordination of the catalytic sites, particularly the types of nitrogen species (e.g., pyridinic N, graphitic N) embedded within the carbon matrix.

Moreover, carbon materials support metal-based catalysts (e.g., Ag, Cu, Sn), which are traditionally active for CO_2RR. The carbon substrate enhances electron transfer, disperses metal particles uniformly, and prevents sintering or detachment during prolonged electrolysis. Porous carbon frameworks with high surface areas and tailored pore structures improve CO_2 adsorption and mass transport, thereby increasing the reaction rate. Emerging research on dual-function electrocatalysts that integrate CO_2 capture and conversion in a single platform has also leveraged functionalized carbon materials as key components. One notable example is the work by Wang et al., who introduced a nitrogen-doped nanoporous carbon–carbon nanotube composite membrane (HNCM-CNT) (Figure 1b-d). This membrane functions as a binder-free, high-performance electrode for the electrocatalytic reduction of CO_2 to formate, achieving a Faradaic efficiency of 81%. The robust structural and electrochemical properties of the membrane also endow it with excellent long-term stability [18].

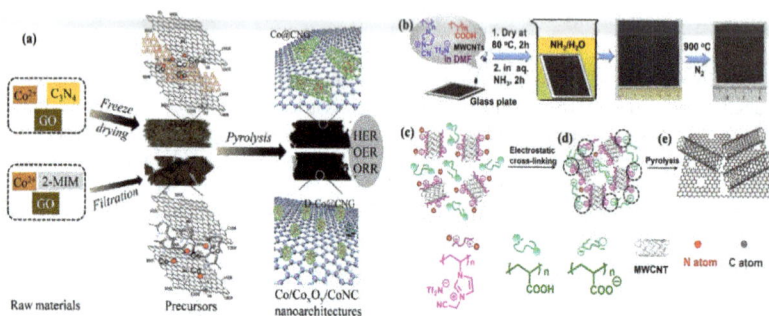

Figure 1. (a) The preparation and structure of Co@CNG and D-Co@CNG trifunctional electrocatalysts. (Reproduced with permission from Ref. [2] Copyright 2017, Royal Society of Chemistry, (b) scheme illustrating the synthetic route to the membranes of the primary PCMVImTf₂N/PAA/CNT film,(c) homogeneous dispersion of CNTs in a solution of PCMVImTf₂N and PAA in DMF, (d) a PCMVImTf₂N/PAA/CNT film, and (e) a HNCM/CNT membrane. (Reproduced with permission from Ref. [18]; Copyright 2017.Wiley-VCH).

Carbonaceous Materials for Energy Storage Systems

Electrochemical devices like batteries, supercapacitors, and hydrogen evolution reaction (HER) technologies rely on the chemistry of electrode and electrocatalyst materials, which are important for energy storage and conversion systems. The chemical bonds existing within these materials determine their abilities in terms of the amount of electrical or chemical energy stored in them. Several factors, such as electrochemical activity, conductivity, and structural stability determine the effectiveness of charge storage and conversion. In general, it can be assumed that the better the electrochemical properties of a material, the higher is its efficiency for storing and using energy. Carbon nanostructures such as graphene, carbon nanotubes (CNTs), and fullerenes are highly important for energy storage devices such as supercapacitors or batteries. Their good applicability in energy storage and transport efficiency is due to their very high electronic conductivity, very large surface area, and strong chemical permanence.

> Graphene-Based Materials: The high electrical conductivity (~10^6 S cm^{-1}), mechanical strength (~1.0 TPa), and high surface area (2630 m² g^{-1}) of graphene make it suitable for use in electrochemical storage. Charge-transport performance can be engineered by a tailored approach that includes heteroatom doping and porous graphene architectures.

> Carbon Nanotubes (CNTs): Structural flexibility, good conductivity (10^2-10^6 S cm^{-1}), and single-walled (SWCNTs) and multi-walled (MWCNTs) forms of CNTs give fast charge transfer in electrodes.

> Fullerenes: Spherical carbon molecules that have excellent charge-transfer capacity, though not much has been studied, have promise for electrochemical applications.

> ➢ Free-Standing Carbon Electrodes: Flexible and lightweight, because of binder-free options such as carbon nanofibers (CNFs), carbon paper (CP), and carbon cloth (CC), these options improve flexibility and performance of electrodes.

Structural engineering strategies like heteroatom doping, porous structures, and 3D frameworks further enhance carbonaceous materials for energy storage devices. Future development will aim at enhancing energy density, scalability, and cycle life for next-generation storage devices.

Advancements in Carbon-Based Electrocatalysts

Nanostructuring and Surface Engineering Strategies: Nanostructuring is a pivotal approach for enhancing the intrinsic catalytic performance of carbon-based materials. By reducing particle size and creating porous architectures, the surface area is significantly increased, leading to a higher density of accessible active sites. Strategies such as templating methods, electrospinning, and soft/hard templating allow precise control over pore structure and morphology. Surface engineering techniques, including heteroatom doping (e.g., nitrogen, sulfur, boron) and defect generation, further modify the electronic structure and surface chemistry of the carbon matrix [19]. For instance, Zhu and co-workers reported an in situ alkaline etching process for fullerene (C_{60}) molecules, which strategically opened up the molecular framework while preserving pentagonal structures, as illustrated in Figure 2a. The aforementioned treatment led to the formation of a product referred to as PD-C, where the pentagonal defects remained exposed and intact. However, these defects imparted markedly enhanced ORR activity compared to the unmodified fullerene [20]. These modifications promote adsorption of reactants and facilitate electron transfer processes, ultimately improving the catalytic efficiency in reactions such as the ORR, HER, and OER.

Synergistic Effects with Metal and Metal-Free Catalysts

Combining carbon-based materials with transition metals (such as Fe, Co, and Ni) or metal oxides can create bifunctional or multifunctional electrocatalysts [20-22]. One such example is when Babu et. al reported the mixed metal–organic frameworks to fabricate CoFe-based hybrid oxyphosphides featuring a highly porous architecture (Figure 2b). This porous framework promotes rapid ion transport and exposes numerous catalytically active sites, enabling the resulting Co_3FePxO material to deliver outstanding performance in OER electrocatalysis [23]. The synergy arises from the interaction between the highly conductive carbon substrate and the catalytically active metal sites. These interactions stabilize the active centres, prevent particle agglomeration, and promote electron transfer across the interface. Another example of a transition metals based catalyst is depicted in Figure 2c. Deng et al. developed a hierarchical nanostructure composed of ultrathin graphene layers (1–3 layers thick) that uniformly encapsulate CoNi nanoalloy particles (CoNi@NC). This structure was synthesized through a bottom-up strategy utilizing Co^{2+}, Ni^{2+}, and $EDTA^{4-}$ as starting materials. The aforesaid architecture delivered enhanced HER activity under acidic conditions [24].

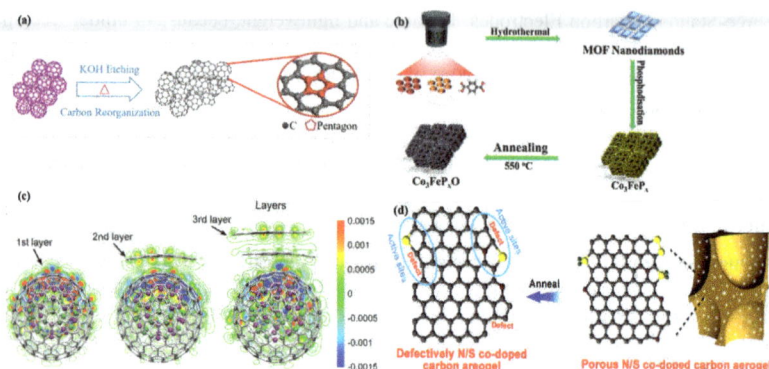

Figure 2. a) Schematic image of synthesizing pentagon defect-rich carbon nanomaterial. (Reproduced with permission from Ref. [20] Copyright 2019, Wiley-VCH, b) Schematic illustration of the synthesis of metal oxyphosphides. (Reproduced with permission from Ref. [23] Copyright 2018, American Chemical Society, c) Diagrammatic representation of a Co–Ni alloy core enclosed within a tri-layer graphene shell. (Reproduced with permission from Ref. [24] Copyright 2015.Wiley-VCH, d) Schematic illustration of the synthesis process of N-Modified S-Defect Carbon Aerogel (Reproduced with permission from Ref. [25] Copyright 2018, Elsevier.

In metal-free systems, dual- or multi-doped carbon materials can exhibit remarkable catalytic properties due to the synergistic effect of various dopants that create local charge density variations and modulate spin distribution. For Instance, Li et al. investigated the synergistic effects of heteroatom co-doping and topological defects by synthesizing a defect-rich carbon aerogel. They synthesized an N/S co-doped carbon aerogel, which was subsequently subjected to high-temperature annealing, thereby partially removing dopant elements while introducing structural defects, as depicted in Figure 2d. The resulting material demonstrated excellent ORR performance under acidic conditions. These findings suggest that the adjacent coordination environment comprising graphitic nitrogen (N), thiophenic sulfur (S), and pentagonal carbon defects (D) acts as a highly active site for acidic ORR The resulting hybrid structures often show enhanced activity, selectivity, and tolerance to reaction intermediates, especially under extreme pH conditions.

Scalable Fabrication Techniques

When moving from laboratory synthesis to industrial-scale production, fabrication methods must be cost-effective, environmentally sustainable, and consistently reproducible. Several techniques are under consideration for large-scale manufacturing, such as spray pyrolysis, chemical vapor deposition (CVD), freeze-drying, hydrothermal carbonization, and direct carbonization of biomass. These methods can be tailored to produce carbon materials with specific porosity, surface chemistry, and structural features. Recent progress in ink formulation combined with improvements in screen-printing technology has made it possible to apply carbon-based catalysts onto flexible substrates, facilitating their integration into practical devices. Minimizing synthesis duration, reducing energy consumption, and ensuring batch-to-batch uniformity are therefore critical factors for scalable production.

Challenges and Future Perspectives

Stability, Durability, and Long-Term Performance Issues: The major barrier that keeps the imagination of using carbon-based electrocatalysts in real applications is their long-term stability under operating conditions. Degradation mechanisms include carbon corrosion, detachment of active sites, and structural collapse owing to long cycling or harsh exposure to electrolytes, and all of them can contribute quite significantly to performance malaise. Redox cycling and mechanical stress also affect the activity loss of the catalysts in fuel cells and water electrolysers. Therefore, the encapsulation of active sites, reinforcement with graphitic frameworks, and optimization of the binder and electrode configurations are being investigated against such potentials.

Sustainable and Scalable Synthesis Approaches: The development of environment- and resource-efficient synthesis routes is a must for implementing carbon-based catalysts widely. Renewable lignin, cellulose, chitosan, and other derivatives from biomass offer sustainability in the fight for carbon precursor replacement. Those biogenic sources are usually inherent with heteroatoms, reducing the need for any further doping steps. Green principles include solventless synthesis, low-temperature processing, and the least waste generated. Also, these processes should be reproducible for uniform catalyst performance in industrial applications.

Integration with Next-Generation Renewable Energy Systems: Carbon-based electrocatalysts are increasingly being considered for integration into advanced renewable energy systems such as metal–air batteries, proton-exchange membrane fuel cells, electrochemical water splitting units, and CO_2 electroreduction setups. Their lightweight nature, tunable surface properties, and corrosion resistance make them attractive candidates for next-generation devices. However, successful integration requires a comprehensive understanding of the catalyst–electrolyte interface, mechanical stability under operating conditions, and compatibility with current collector materials. The design of multifunctional electrode architectures that incorporate carbon-based materials is key to achieving efficient and stable performance in real-world applications.

Conclusion

In the context of a rapidly growing expanse in energy conversion and storage technology, the demand for electrocatalysts that are cost-effective, efficient, and sustainable becomes more relevant. Carbon-based materials, and particularly those that are metal-free and heteroatom-doped, have emerged as great alternatives to classical noble metal catalysts. Their advantageous properties-large surface area, tunable electronic structure, chemical stability, and environmental benignity-make carbon-based materials prime candidates for a broad spectrum of electrocatalytic applications, including fuel cells, metal-air batteries, and water splitting. This chapter covered major advances in the synthesis, functionalization, and practical applications of carbon-based electrocatalysts, focusing on two-dimensional structures. Heteroatom doping, defect engineering, and nano-structuring have been proven to be effective in boosting catalytic performance by tuning active sites and enhancing charge transfer. Increasing efforts in understanding the correlations of structure to activity and the mechanistic support of carbon frameworks are all paving reliable pathways for rational design toward next-generation catalysts.

Challenges still remain, particularly in the realms of long-term stability of the catalysts, controlled distribution of active sites, and methods of scaling for synthesizing carbon-based electrocatalysts. Collaborative efforts will be needed across material science, electrochemistry, and engineering to address these challenges. Future directions of research should be set on eco-friendly, scalable fabrication routes, advanced characterization efforts for deeper mechanistic understanding, and effective integration of carbon-based catalysts into commercial energy devices. In summary, carbon-based materials present ample opportunities for the advancement of sustainable electrocatalysis. Continued strides in this field will be paramount in servicing the need for clean and renewable energy all over the world.

References

[1] Anandhababu, G., Huang, Y., Babu, D. D., Wu, M., & Wang, Y. (2019). Oriented growth of ZIF-67 to derive 2D porous CoPO nanosheets for electrochemical-/photovoltage-driven overall water splitting. Advanced Functional Materials, 28(9), 1706120.. https://doi.org/10.1002/adfm.201706120

[2] Huang, Y., Liu, Q., Lv, J., Babu, D. D., Wang, W., Wu, M., Yuan, D., & Wang, Y. (2017). Co-intercalation of multiple active units into graphene by pyrolysis of hydrogen-bonded precursors for zinc-air batteries and water splitting. Journal of Materials Chemistry A, 5(39), 20882-20891.. https://doi.org/10.1039/C7TA06677E

[3] Novoselov, K. S., et al. (2004). Electric field effect in atomically thin carbon films. Science, 306(5696), 666-669.. https://doi.org/10.1126/science.1102896

[4] Dai, L., Xue, Y., Qu, L., Choi, H.-J., & Baek, J.-B. (2016). Metal-free catalysts for oxygen reduction reaction. Chemical Reviews, 115(11), 4823-4892.. https://doi.org/10.1021/cr5003563

[5] Li, Y., Li, Y., Zhu, E., & Zhang, H. (2016). Heteroatom-doped carbon materials for advanced electrochemical energy storage. Advanced Energy Materials, 6(9), 1600751.. https://doi.org/10.1002/aenm.201600751

[6] Qu, L., Liu, Y., Baek, J.-B., & Dai, L. (2015). Nitrogen-doped graphene as efficient metal-free electrocatalyst for oxygen reduction in fuel cells. ACS Nano, 4(3), 1321-1326.. https://doi.org/10.1021/nn901850u

[7] Shui, J., Wang, M., Du, F., & Dai, L. (2015). N-doped carbon nanomaterials are durable catalysts for oxygen reduction and oxygen evolution reactions. Nature Communications, 6, 6392.. https://doi.org/10.1126/sciadv.1400129

[8] Gong, K., Du, F., Xia, Z., Durstock, M., & Dai, L. (2009). Nitrogen-doped carbon nanotube arrays with high electrocatalytic activity for oxygen reduction. Science, 323(5915), 760-764. https://doi.org/10.1126/science.1168049. https://doi.org/10.1126/science.1168049

[9] Zhao, Y., Watanabe, K., & Hashimoto, K. (2015). Self-supporting oxygen reduction electrocatalysts made from a nitrogen-rich network polymer. Journal of the American Chemical Society, 134(48), 19528-19531.. https://doi.org/10.1021/ja3085934

[10] Wang, X., Jia, Y., Mao, X., Zheng, Y., & Qiao, S.-Z. (2017). Graphene-based electrocatalysts for hydrogen evolution reaction. ACS Catalysis, 7(1), 211-229.

[11] Huang, Y., Babu, D. D., Peng, Z., et al. (2020). Atomic modulation, structural design, and systematic optimization for efficient electrochemical nitrogen reduction. Advanced Science, 7(4), 1902390.. https://doi.org/10.1002/advs.201902390

[12] Huang, Y., Babu, D. D., Wu, M., & Wang, Y. (2018). Synergistic supports beyond carbon black for polymer electrolyte fuel cell anodes. ChemCatChem, 10(20), 4495-4513.. https://doi.org/10.1002/cctc.201801094

[13] Yang, S., Feng, X., Wang, X., & Müllen, K. (2013). Graphene-based carbon nitride nanosheets as efficient metal-free electrocatalysts for oxygen reduction reactions. Angewandte Chemie International Edition, 50(22), 5339-5343.. https://doi.org/10.1002/anie.201100170

[14] Liu, X., Dai, L., & Chen, Z. (2019). Functional carbon-based nanomaterials for electrocatalytic energy conversion. Chemical Society Reviews, 48(10), 3072-3101.

[15] Peng, H., et al. (2013). High performance Fe- and N-doped carbon catalyst with graphene structure for oxygen reduction. Applied Catalysis B:Environmental, 130-131, 31-38.. https://doi.org/10.1038/srep01765

[16] Han, X., Ling, X., Yu, D., Xie, D., Li, L., Peng, S., Zhong, C., Zhao, N., Deng, Y., & Hu, W. (2019). Atomically dispersed binary Co-Ni sites in nitrogen-doped hollow carbon nanocubes for reversible oxygen reduction and evolution. Advanced Materials, 31(49), 1905622.. https://doi.org/10.1002/adma.201905622

[17] Liu, X., Dai, L., & Chen, Z. (2016). Metal-free catalysts for electrochemical CO_2 reduction. Nature Communications, 7, 10921.

[18] Wang, H., Jia, J., Song, P., Wang, Q., Li, D., Min, S., Qian, C., Wang, L., Li, Y. F., Ma, C., Wu, T., Yuan, J., Antonietti, M., & Ozin, G. A. (2017). Efficient electrocatalytic reduction of CO_2 by nitrogen-doped nanoporous carbon/carbon nanotube membranes: A step towards the electrochemical CO_2 refinery. Angewandte Chemie International Edition, 56(27), 7847-7852.. https://doi.org/10.1002/anie.201703720

[19] Abbas, S. C., Babu, D. D., Anandhababu, G., Wu, M., & Wang, Y. (2018). Novel N-Mo_2C active sites for efficient solar-to-hydrogen generation. ChemElectroChem, 5(8), 1186-1190.. https://doi.org/10.1002/celc.201701365

[20] Zhu, J., Huang, Y., Mei, W., Zhao, C., Zhang, C., Zhang, J., Amiinu, I. S., & Mu, S. (2019). Effects of intrinsic pentagon defects on electrochemical reactivity of carbon nanomaterials. Angewandte Chemie International Edition, 58(12), 3859-3864.. https://doi.org/10.1002/anie.201813805

[21] Wang, W., Babu, D. D., Huang, Y., Lv, J., Wang, Y., & Wu, M. (2018). Atomic dispersion of Fe/Co/N on graphene by ball-milling for efficient oxygen evolution reaction. International Journal of Hydrogen Energy, 43(22), 10351-10358.. https://doi.org/10.1016/j.ijhydene.2018.04.108

[22] Anandhababu, G., Manimuthu, V., Baskaran, S., Babu, D. D., Wu, M., & Wang, Y. (2021). Profuse surface activation of Ir-dispersed titanium nitride bifunctional electrocatalysts. Advanced Energy and Sustainability Research, 2(2), 2000054.. https://doi.org/10.1002/aesr.202000054

[23] Babu, D. D., Huang, Y., Anandhababu, G., Ghausi, M. A., & Wang, Y. (2017). Mixed-metal-organic framework self-template synthesis of porous hybrid oxyphosphides for efficient

oxygen evolution reaction. ACS Applied Materials & Interfaces, 9(44), 38621-38628..
https://doi.org/10.1021/acsami.7b13359

[24] Deng, J., Ren, P., Deng, D., & Bao, X. (2015). Enhanced electron penetration through an
ultrathin graphene layer for highly efficient catalysis of the hydrogen evolution reaction.
Angewandte Chemie International Edition, 54(7), 2100-2104..
https://doi.org/10.1002/anie.201409524

[25] Li, D., Jia, Y., Chang, G., Chen, J., Liu, H., Wang, J., Hu, Y., Xia, Y., Yang, D., & Yao, X.
(2018). A defect-driven metal-free electrocatalyst for oxygen reduction in acidic electrolyte.
Chem, 4(10), 2345-2356.. https://doi.org/10.1016/j.chempr.2018.07.005

Electrocatalysts and Advanced Materials for Sustainable Energy Storage   Materials Research Forum LLC
Materials Research Foundations 182 (2025) 135-159   https://doi.org/10.21741/9781644903797-10

Chapter 10

Stability and Durability of Graphene-based Materials for Supercapacitor Applications: Challenges and Solutions

HARITHA Valiyaveettil Padi[1,a], BINITHA N. Narayanan[1,2,b*]

[1]Department of Chemistry, University of Calicut, Calicut University (P.O.)-673635, Kerala, India

[2]Inter University Centre for Hydrogen & Energy Storage, University of Calicut, Kerala, India

[a]harithavp2014@gmail.com, [b]binitha@uoc.ac.in,

Abstract

Supercapacitors are vital for next-generation energy storage due to their high-power density, rapid charge-discharge capability, and long cycle life. Graphene-based electrodes are well-suited for supercapacitors due to their exceptional conductivity and large surface area. However, several challenges hinder long-term durability; pristine graphene restacks, thereby reducing ion accessibility, while reduced graphene oxide (rGO), though cost-effective, has structural defects, low surface area, and lower conductivity. High-quality graphene obtained by chemical vapor deposition offers high conductivity but is costly and lacks functionalities that aid nanocomposite formation which is necessary for improved performance. One solution to these issues lies in the edge functionalization of graphene via various mechanochemical graphite exfoliation methods, particularly ball milling, which introduces active sites without affecting the in-plane π-conjugation, retaining conductivity, and also preventing layer restacking. Structural modifications, such as heteroatom doping, porosity introduction, development of 3D architectures, nanocomposite formation, etc., enhance the energy density of high-power-density graphene-based electrodes. Additionally, integrating graphene with other conductive layered materials improves electrochemical activity and stability, delivering excellent supercapacitor performance. Together, these strategies improve the durability and stability of supercapacitors, paving the way for cost-effective, high-performance, and high-quality graphene-based energy storage systems. This chapter explores the advantages and current challenges associated with graphene-based supercapacitors while examining potential solutions to overcome the limitations.

Keywords

Graphene, Supercapacitor, Stability & Durability, Graphite Exfoliation, Edge-Functionalization

Contents

Introduction

The emergence of alternatives for energy conversion and their storage is consequent upon the accepted fact that depletion of conventional energy sources will lead to the biggest worldwide energy scarcity in the future [1,2]. Storage difficulty is the major challenge in the case of renewable energy sources with intermittent supply [3]. The development of efficient energy storage devices is thus highly required for sustainable energy in the future [4]. Studies have proved that the energy requirement will be approximately double the current consumption by the end of this century due to rapid human population growth and advancement of technology [5]. So, the novel energy storage technologies should be scalable, clean, cost-effective, eco-friendly, and sustainable to rely completely on them [6–8]. Among dominant energy storage devices, supercapacitors bring together the key properties of conventional rechargeable batteries and capacitors [7]. Supercapacitors possess higher energy density than capacitors and power density far better than rechargeable batteries (Figure 1) [9]. Energy storage devices like lithium-ion batteries (LIBs) are capable enough in terms of energy density; but they fail to deliver quick power for applications like emerging electric heavy vehicles and other electronic devices [4]. Further, their limited cycle life is another hurdle that will lead to extensive pollution by disposal in these growing ecological concerns [10]. The electrochemical reactions in batteries lead to degradation of the electrode material over time, which shortens their life [11]. The rapid replacement of batteries within a short period embodies the accumulation of unavoidable electronic waste. Rather than batteries, fuel cells, electrochemical capacitors, dielectric capacitors, etc., can also be employed for energy storage purposes, and among them, capacitors display faster charge-discharge rates along with reduced emission [12]. In the case of capacitors, lower energy density constraints the flexibility in various applications, like in the miniaturization of electronic gadgets [7]. These shortcomings of established energy storage technologies shift the focus towards supercapacitors, also known as ultracapacitors, which combine the best of both capacitors and batteries, like reasonable energy density and outstanding power density [8]. Supercapacitors stand out for their exceptional cycling life (>100,000 cycles), far surpassing the limited lifespan of batteries (<1,000 cycles), making them a more sustainable option for emerging green technologies [11].

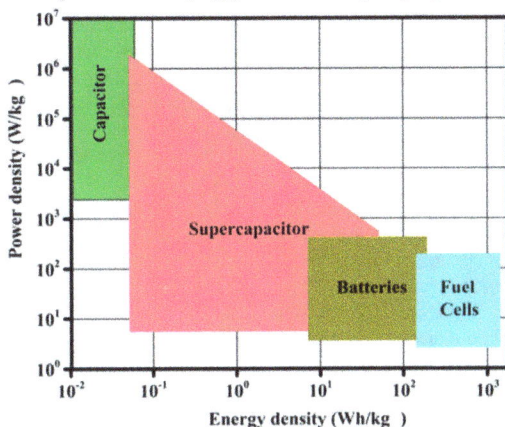

Figure 1. Ragone plot of different energy storage devices (Reproduced with permission from Ref. [9], Copyright 2020, Multidisciplinary Digital Publishing Institute).

Supercapacitors, having less charging time, high energy storage capability, good energy density, exceptional power density, performance retention over a large number of continuous usages, etc., are thus a recently developing technology to substitute other mainstream energy storage devices with certain limitations [13]. Materials with high active surface area, excellent thermal and chemical stability, non-corroding nature, and high electrical conductivity are often the preferred electrodes in supercapacitors [14–16]. So, recent research emphasizes developing advanced electrode materials that can be produced in a low-cost, scalable manner following green principles. Among carbon-based materials, graphene, carbon nanotubes, and activated carbon are frequently used as electrode materials in supercapacitors [17]. The choice of electrodes and electrolytes has a great impact on the development of stable and durable supercapacitors [18]. While focusing on efficient fabrication techniques, the option of current collectors also influences the overall performance of supercapacitors [19].

This chapter discusses the role of supercapacitors in future energy storage and the various strategies to develop more efficient supercapacitors. Here, we mainly focus on the importance of the choice of electrode materials and the exceptional performance of graphene as an electrode material in supercapacitors. The role of green preparative pathways of graphene and their quality compared to prevalent graphene oxide (GO)-based graphene is also briefed. The major challenges regarding the stability and durability of graphene-based supercapacitors and remedies to overcome their shortcomings are also narrated here.

Importance of Supercapacitors in Energy Storage Applications

The Leyden Jar was the initial form of the currently used supercapacitors. Several modifications in this basic form have led to the modern form of supercapacitors [14,20]. Two electrodes separated by an ion-permeable membrane containing a suitable electrolyte constitute the current form of a supercapacitor device [21–23]. The forthcoming era of supercapacitors is acclaimed for its compatibility in a vast number of applications. Currently, their integration into electric and hybrid vehicles constitutes a primary area of investigation among researchers. The ability for quick power bursts enables fuel efficiency, and there are no emissions compared to conventional fuel vehicles [4]. Supercapacitors also find applications in various portable smart electronic gadgets like mobile phones, laptops, watches, digital cameras, etc [24,25]. Additionally, in this advanced era of technology, all electronic gadgets are being designed to be more compact and flexible, while maintaining efficiency. To support this growth, the core elements, like the energy storage part, need to be compact and extremely flexible [26]. Supercapacitors can be made incredibly flexible, relying on printable microelectrodes with higher specific capacitance, and are designed to ensure maximum efficiency, unaffected by bends or twists [27]. Most flexible supercapacitors use versatile forms of carbon materials as electrode materials and fabric-based current collectors [28–30]. There are different types of supercapacitors, like coin cell supercapacitors, cylindrical supercapacitors, axial supercapacitors, flexible supercapacitors, pouch cell supercapacitors, and micro supercapacitors, and their implementation depends on what the application demands [31–33]. Supercapacitors have grown to be an indispensable energy storage option over time due to all their superior features over established energy storage methods [21]. They can also be interconnected with renewable energy sources, as they could help in power stabilization, and this paves the way for new possibilities in the future [4,34]. Table 1 shows how the performance of supercapacitors leads to their exploitation for various applications in comparison to capacitors and batteries.

Table 1. Comparison of the performance of supercapacitors with batteries and capacitors

Function	Battery	Capacitor	Supercapacitor
Charging time	Hours	μs-ms	ms-min
Discharging time	1-900 min	μs-ms	ms-days
Energy density	100-200 Wh/Kg	< 0.1 Wh/Kg	1-10 Wh/Kg
Power density	1000-3000 W/Kg	> 10000 W/Kg	500-10000 W/Kg
Cycle life	200-1000	10^6-10^8	10^6-10^8

With regard to real-sense implementation, the cycle life of supercapacitors is a key parameter since this technology is being implemented to lower waste generation from electronic gadgets compared to batteries [10]. Safety concerns also reinforce this development. The working temperature range of the supercapacitors goes below -40°C, which is unattainable for batteries or capacitors [35]. Another greater advantage regarding supercapacitors is the possibility of supercapattery systems obtained by merging supercapacitors and battery electrodes [36]. Thus, they can be considered an excellent alternative to the current commercially used energy storage systems.

Electrocatalysts and Advanced Materials for Sustainable Energy Storage   Materials Research Forum LLC
Materials Research Foundations 182 (2025) 135-159   https://doi.org/10.21741/9781644903797-10

Energy Storage Mechanisms in Supercapacitors

The properties of electrode materials and the nature of electrolytes determine the operating mechanism of energy storage unique to each combination [7]. Charging induces movement of electrons from the positive to the negative electrode, and sequentially, ions of electrolyte move towards oppositely charged electrodes [35]. In order to maintain the external charge balance, either the formation of an electrical double layer (EDL) or the sparking of Faradaic reactions occurs [37]. This process occurs reversibly during the discharge cycle. Under the charge storage mechanism in supercapacitors, they can be categorized into three types: electrical double layer capacitors (EDLC), pseudocapacitors (PC), and hybrid capacitors [8,38]. The pictorial representation of their respective charge storage mechanism is given in Figure 2. This classification is also grounded in the electrode materials used and the electrochemical properties of so supercapacitors fabricated [7]. A physical charge separation assists energy storage in EDLC rather than redox reaction-mediated energy storage in PC [39]. Asymmetric/hybrid capacitors have an interplay of both mechanisms [40]. The basic formula for energy storage in capacitors can be directly applied in the case of supercapacitors, which is given as equation (1), in which all the components are directly connected to properties that enhance the performance of supercapacitors [7].

$$C = \frac{\varepsilon_0 \varepsilon_r A}{d} \tag{1}$$

Here, ε_0 is the permittivity of free space and ε_r is the relative permittivity of the dielectric material, A is the electrode surface area, and d represents the distance between electrodes. It is clear from the equation that tuning the area of electrodes and interplanar thickness can enhance the overall performance of supercapacitors [7]. Also, the need of more focused research towards developing high surface area electrode materials derived from graphene-like systems is evident [41,42].

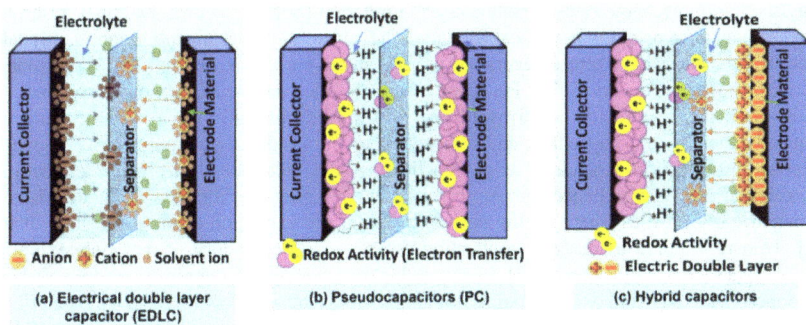

Figure 2. Schematic representation of energy storage mechanism in different types of supercapacitors (a) EDLC, (b) Pseudocapacitor, and (c) Hybrid capacitor (Reproduced with permission from Ref. [43], Copyright 2022, Multidisciplinary Digital Publishing Institute)

Electrocatalysts and Advanced Materials for Sustainable Energy Storage   Materials Research Forum LLC
Materials Research Foundations 182 (2025) 135-159   https://doi.org/10.21741/9781644903797-10

Electrical double layer capacitors (EDLC): The application of voltage across a supercapacitor initiates the forced movement of electrons and accumulation of ions of electrolytes at the interface between inversely charged electrode and electrolyte as mentioned [44]. This is referred to as an EDL, and it serves as a depot for energy storage where energy is stored non-Faradaically [45,46]. Since the mechanism holds the formation of an EDL, high surface area materials are greatly preferred when designing EDLC electrodes, and the layer thickness is supposed to be maintained at a minimum [47]. The formation of double layer can be discussed concerning three proposed models. Helmholtz's (1853) model proposed charge accumulation over electrodes by an adsorption-desorption mechanism as a result of the movement of ions [35,48]. Additionally, the Gouy and Chapman model proposed the formation of an EDL involving the distribution of countercharged ions of the electrolyte [15]. Stern's model refined the concept by adding the idea of a Stern layer and a diffused layer, leading to specific ion adsorption and charge distribution along with inhibition of ion recombination [49]. The potential difference between the inner layer and outer layer, called the zeta potential, describes the degree of charge storage in EDLC [15,50]. EDLC materials offer the potential to significantly increase the power density [51]. As energy storage does not involve any reactions, the life of EDLC electrodes can be substantially better [52]. Consequently, supercapacitors can withstand continuous usage over 100000 cycles. Materials having high electrical conductivity, excellent accessible surface area, and excellent chemical or thermal stability are often under this category of electrode materials [53,54]. Most of the nanostructured carbonaceous materials mainly follow the EDLC mechanism of energy storage [55].

Pseudocapacitors (PC): In pseudocapacitors, energy storage is driven by charge transfer between the electrode and electrolyte through redox reactions [56]. Here, the interaction between the electrode and electrolyte results in the formation of chemical species aided by the electron transfer across the double layer within the system [57]. This charge transfer occurs on the surface as well as on the immediate bulk, and hence, PCs can achieve comparably higher specific capacitance and energy density than EDLC [58]. Thus, PC exhibits a Faradaic charge storage mechanism, their sluggish charge-discharge rates persist as a key limitation [59]. PC can follow either adsorption pseudocapacitance or redox pseudocapacitance. The adsorption mechanism predominantly takes place in 2D materials, where a reversible process involving the adsorption and desorption of hydrogen molecules occurs [7,56]. For those materials following the adsorption mechanism, CV profiles are more symmetric for forward and backward scans [56]. The redox mechanism is frequently seen in supercapacitors, and it stems either from redox functionalities over electrode materials or diffusion-dependent redox reactions at the electrode-electrolyte interface [60]. Intercalation pseudocapacitance can also be observed in supercapacitors and is characterized by asymmetric CV profiles where continuous ion insertion and extraction occur at varying voltages [61]. Transition metal oxides and conducting polymers are a category of materials exhibiting pseudocapacitive energy storage [62]. Metal oxides for supercapacitor applications are now being developed as mixed metal oxides ranging from binary to quaternary composites [63]. The primary constraint of pseudocapacitive materials is their limited cycle life in comparison with EDLC [35].

Hybrid supercapacitors: This class of supercapacitors aims at a synergistic combination of EDLC materials and pseudocapacitive or battery-like materials to attain the highest specific capacitance possible and to compensate for the shortcomings of the component materials [64]. Hybrid supercapacitors are being developed by combining a graphene-like highly conductive matrix with various metal oxides, metal sulphides and conducting polymers etc., to constitute hybrid electrode materials [65]. Tremendous improvement in energy density is observed in many cases, uninfluencing the power density, and their cycle life are also found to be increasing [65]. Thus, the emerging trend in supercapacitor electrode material development can be considered to benefit the improvisation in future technology. Thus, integration of both Faradaic and non-Faradaic energy storage materials is the major characteristic of hybrid supercapacitors [66]. An asymmetric supercapacitor and supercapattery also falls under this category [67]. In an asymmetric supercapacitor, usually one electrode is made up of carbonaceous EDLC materials, and the other by a pseudocapacitive material. It has higher energy density and power density than that obtained upon their individual exploitation [64]. Asymmetric devices provide the added advantage of an extended potential window especially while dealing with aqueous electrolytes [68]. Supercapattery conjoins a supercapacitor electrode with a battery electrode which in turn manifests the properties of both supercapacitors and batteries [36].

Graphene

Graphene is the two-dimensional carbon allotrope composed of sp^2-hybridized carbon atoms arranged in a hexagonal lattice which forms a sheet-like structure with a thickness of just one atom in a single-layer structure (Figure 3a) [69]. It can be regarded as the basic constituent of all other nano-dimensional materials of carbon, like 0D fullerene, 1D carbon nanotubes (CNTs), and 3D graphite, as the honeycomb network of 2D graphene can be converted to all the other dimensions [70]. The properties of graphene are largely determined by the $2p_x$ orbitals exhibiting π-symmetry orientation [71]. Graphene possesses a theoretical surface area of $2630 \, m^2/g$, a specific capacitance of $550 \, F/g$, a tunable band gap, high electrical conductivity, and electron transfer properties [72]. Due to its exceptional electrical conductivity and high surface area, it is often employed as an electrode material in supercapacitors [70]. As energy is stored electrostatically in graphene, it will diminish electrode degradation and thereby increase the cycling life of supercapacitors [10]. Hence, graphene-based materials are greatly admired for fabricating long-life, high-capacitive electrodes. The utilization of graphene and its nanocomposites can also foster the replacement of metal-based electrodes, which mostly produce less durable supercapacitors [73].

Synthetic Routes of Graphene and Their Influence on Its Properties

There are various top-down and bottom-up methods for the preparation of graphene. Among them, micromechanical exfoliation, chemical exfoliation, electrochemical exfoliation, liquid phase exfoliation, arc discharge, chemical vapour deposition (CVD), and epitaxial growth are the widely used methods [74]. The different methods are mentioned in the following sections and represented in Figure 3b.

Micromechanical exfoliation: Scotch tape-assisted micromechanical exfoliation of graphite to graphene introduced by Geim and co-workers in 2004 is the first straightforward method to graphene preparation that lead to the Nobel in the year 2010, where the weak van der Waals' attractive forces between graphite layers were resolved by physical methods [80]. Here, the continuous application of force by the tape peeled off graphite up to a single layer of graphene, and the quality of so-produced graphene is controlled by the exerted external force [81]. Graphene obtained in this method displays a high surface area and excellent electrical conductivity [82]. However, this method is unsuitable for scalable preparation and has limitations related to the production of functionalized graphene having more advanced properties [83]. With the aim of getting a better yield of graphene, Jayasena et al. developed a lathe-like equipment to produce graphene from highly ordered pyrolytic graphene (HOPG). Here, HOPG immersed in a resin medium is split using an extremely sharp diamond wedge to obtain graphene sheets of a thickness of tens of nanometres [84]. A modified version of the Scotch-tape method was introduced by J. Chen and co-workers employing a three-roll mill machine, where an adhesive, that is a mix of polyvinyl chloride (PVC) and dioctyl phthalate (DOP), along with graphite, is fed to the equipment [80]. The difficulty in removing the adhesive is the major challenge here, rather than the scalability. All these methods are inadequate to give functionalized graphene having high processability and advanced properties [83].

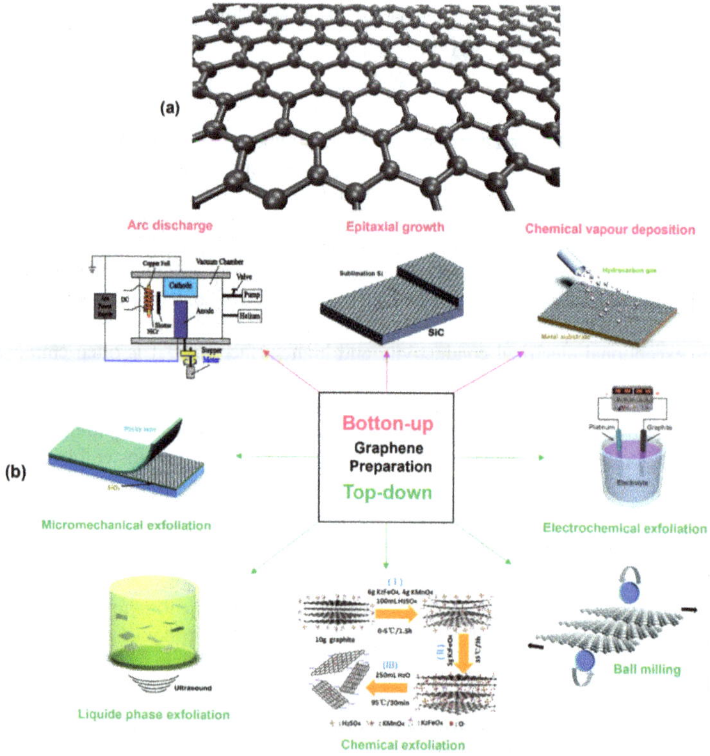

Figure 3. (a) Ball and stick model of a portion of graphene sheet (Reproduced with permission from Ref. [75], Copyright 2019, Multidisciplinary Digital Publishing Institute) (b) Top-down and bottom-up methods of graphene preparation (Reproduced with permissions from Ref. [76], Copyright 2012, Elsevier)

Electrocatalysts and Advanced Materials for Sustainable Energy Storage   Materials Research Forum LLC
Materials Research Foundations 182 (2025) 135-159   https://doi.org/10.21741/9781644903797-10

Chemical exfoliation: This can be considered the widely accepted method to synthesize graphene in a scalable way. In chemical oxidation, graphite is converted to graphene oxide (GO) in a cost-efficient manner using strong oxidising agents [85]. Among the oxidation routes, Hummer's method and its modified versions are prominent in the scientific community. In Hummer's method, $KMnO_4$ is the oxidising agent, which only performs partial oxidation and hence has a lower yield, which leads to various modified forms of this baseline procedure [86]. A modified version of this uses a mixture of H_2SO_4, $KMnO_4$, $NaNO_3$, and H_2O_2 for producing graphite oxide, which, upon ultrasonication, gives GO [87]. This route is considered non-ecofriendly as toxic NO_x and SO_x gases are emitted as byproducts [86]. The graphene so obtained is rich with oxygen functionalities and hence can be easily modified to nanocomposites with various materials and molecules [88]. However, the obtained graphene suffers from broken aromatic π-conjugation due to drastic oxidising conditions, leading to conductivity loss and hence may have limited performance as supercapacitor electrode [89].

Electrochemical exfoliation: For the electrochemical exfoliation of graphite, two or three electrodes can be used, where graphite is one electrode and the other can be copper or platinum [90]. These electrodes, immersed in an aqueous or non-aqueous medium, constitute the setup up and then upon applying an electric current, graphite is expanded [91]. To the expanded graphite, electrolyte ions are entered in between the layers, causing an increase in the interlayer spacing, resulting in the formation of graphene [90]. Graphite can either act as a cathode or an anode during electrochemical exfoliation. Commonly used electrolytes for this purpose are NaOH, H_2SO_4, Na_2SO_4, NH_4Cl, H_2O_2, K_2SO_4, etc., and the choice of electrolyte determines the degree of exfoliation of graphite [92]. Electrochemically exfoliated graphene can be of high purity, oxidisable, and less defective. This method is also suitable for producing functionalized graphene having improved dispersibility and better exfoliation [90]. This method can be combined with thermal exfoliation or ultrasonication to improve the properties of the obtained graphene. In most cases, the resulting graphene consists of a few layers [38,91].

Liquid phase exfoliation (LPE): LPE involves the exfoliation of graphite by intercalating organic solvents or surfactants between its layers, facilitated by ultrasonication [93]. The resulting shear forces surmount the van der Waals attractions holding the graphite layers together, leading to their separation[93]. Generally, solvents whose surface energies are comparable to that of graphene are employed for this purpose [94]. The charge transfer between a suitable solvent and graphite, where they act as donor or acceptor, enables efficient exfoliation of graphite [93]. Some commonly used organic solvents include 1-methyl-2-pyrrolidinone (NMP), N,N-dimethylformamide (DMF), N,N-dimethylacetamide (DMA), etc [95]. LPE produces less-defective graphene in suitable solvents through sheer force exfoliation; however, the method still suffers from low yield [96]. The toxicity of ideal organic solvents and the difficulty in the removal of surfactants are the major challenges regarding LPE while acquiring pure graphene [97]. The possibility of graphene restacking, the need for long-time sonication, and chances for surface destruction of graphene during surfactant or solvent removal can also hinder their impactful employment in large-scale graphene synthesis [97]. A mixture of green solvents like isopropyl alcohol (IPA) with water or IPA with ethyl alcohol or IPA with acetophenone, is also employed for this purpose [98].

Arc discharge: A controlled production of graphene nanosheets can be attained by the arc discharge method, where current is allowed to pass between highly pure graphite electrodes in the presence of H_2 or He gas, producing highly crystalline material [99]. Hydrogen gas can prevent the rolling up of graphene sheets by breaking the bonds between carbon atoms [80,100]. There are various modified forms of this method to synthesize graphene having desirable properties, as the conventional method results in graphene sheets that are clustered, defective, and small in size [80]. Even though, its highly conducting and scalable nature promote applications in energy storage.

Chemical vapour deposition (CVD): CVD can be adapted for lab-scale graphene preparation where a volatile carbon source is used to deposit graphene over metal substrates like Pt, Ni, Cu, Pd, etc., under inert, high-temperature, high-vacuum conditions [99]. Methane, ethane, acetylene, methanol, and ethanol are some commonly used precursors for graphene deposition by this method [99]. Typically, vapours of these precursors are deposited over metal substrates inside a high-temperature horizontal hot-walled tubular furnace to obtain single to few-layer graphene, depending upon the efficiency of the employed method [101]. This method is comparatively costly, low-yielding, and hence unsuitable for bulk graphene production [101].

Epitaxial growth: This method aims at a more controllable manner of graphene production over various substrate surfaces using atomic carbon precursors [102]. The growth of graphene layers over SiC is a prevalent method using resistive heating or electron beam heating to decompose SiC [102]. An ultrahigh vacuum environment is preferred here to avoid possible contamination and hence gives superior performance when utilized for electronic applications, like in supercapacitors [103]. The expense of the method limits its usage for industrially producing graphene [99].

Ball-mill exfoliation of graphite: A significantly more robust ball mill-assisted mechanochemical method for exfoliating graphite into graphene has been recently established, offering a promising route to high-quality, edge-functionalized graphene through green routes [104]. During ball milling, graphite is mixed with a selected milling agent at an optimized ratio, and the mixture is milled using stainless steel or zirconia balls in a stainless steel container [105]. The stronger shear force arising from the hitting of the balls of the mill helps to exfoliate graphite layers, and molecules of milling media intercalate between layers to ease the process [104]. The high-energy balls could also break the C-C bonds, facilitating edge functionalization or heteroatom doping, which improves the electrochemical properties of graphene [80]. The milling agents used can be carbohydrates, polyaromatic hydrocarbons, N-rich compounds, etc., which can be chosen according to the targeted functionality or dopants over graphene [106]. As the basal plane functionalization is absent, aromatic conducting π-network can be preserved, resulting in superior conductivity comparable with that of pristine graphene obtained by various sophisticated bottom-up methods, but on an industrial scale milder temperature conditions, which signifies their promising nature [89]. The interaction of the milling agent with graphite determines the exfoliating mechanism [104]. Dry and wet milling methods are used with various milling agents like melamine, urea, sugar, jaggery, dry ice, NaCl, NH_4Cl, naphthalene, etc., most of which result in less-defective graphene [104,106,107].

Graphene-based Electrode Materials and their Advantages in Supercapacitors

Graphene can be composited with metal oxides, metal sulphides, conducting polymers, etc. in order to attain improved supercapacitor performance. A broad range of graphene-based materials, including modified graphene, graphene-metal oxide nanocomposites, graphene-conducting polymer nanocomposites, graphene-other carbon nanocomposites, and, as well as other emerging graphene nanocomposites, have gained significant attention for their application in supercapacitor electrode fabrication (Figure 4) [108].

Electrocatalysts and Advanced Materials for Sustainable Energy Storage   Materials Research Forum LLC
Materials Research Foundations 182 (2025) 135-159   https://doi.org/10.21741/9781644903797-10

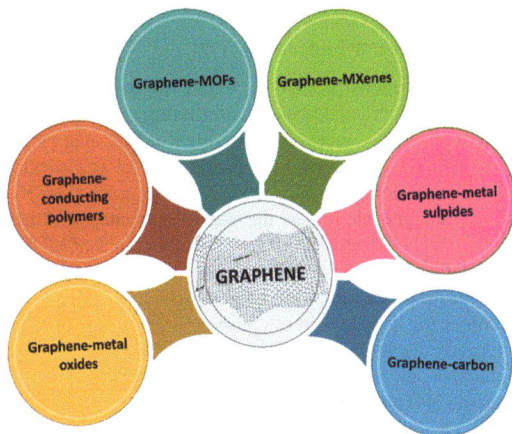

Figure 4. Graphene-based electrode materials for supercapacitor applications

Graphene-Metal Oxide Nanocomposites: Graphene-metal oxide nanocomposites exhibit enhanced specific capacitance and outstanding cyclic stability, owing to the synergistic interaction between the constituent materials [108]. Nanocomposites of graphene with oxides of Mn, Ru, Ni, Co, Cu, Zn, Fe, Mo, Sn, Zr, etc. are renowned for their supercapacitor performance, and among those that are non-toxic and have low-cost preparative methods are mostly preferred to employ in supercapacitor electrodes [109]. Among several metal oxides, RuO_2 was the firstly used transition metal oxide (TMO) in energy storage applications [69]. However, efforts to identify metal oxides with lower toxicity, low-cost and greater abundance have led to the adoption of compounds such as MnO_2, ZnO, Fe_2O_3 and Co_3O_4. Their combination with graphene enhances their surface area, electrical conductivity of the combination, and minimizes the charge transfer resistance between metal oxides and the current collector [109]. Mixed metal oxides and their nanocomposites with graphene are an emerging group of materials within this domain. Metal oxides also impart a notable pseudocapacitive nature to supercapacitor electrodes resulting in improvised energy density [69].

Graphene-Conducting Polymer Nanocomposites: Organic conductors like conducting polymers are known for their pseudocapacitive energy storage capability due to fast and reversible redox reactions endorsed by the π-conjugation [70]. Simple and low-cost synthetic methods, high specific capacitance, environmental compatibility, etc., also make them attractive for energy storage applications. Among conducting polymers, polyaniline (PANI), polythiophene (PTh), polypyrrole (PPy), etc., and their derivatives are most common in supercapacitor electrodes [108]. However, the major hindrance of conducting polymers is their low cyclic stability due to continuous shrinking and swelling of the polymer during charge-discharge cycles, leading to self-degradation [110]. Graphene-conducting polymer nanocomposites are often favoured for their enhanced durability and longer cyclic life compared to conducting polymers on their own [70]. Graphene serves as the active site for polymer chain propagation and exhibits an enhanced electron transfer pathway in these nanocomposites [111]. This combination also helps to prevent potential graphene restacking and the self-degradation of conducting polymers. Improved energy and power densities are the advantages of their hybrid material [72].

Graphene-Metal Organic Framework (MOF) Nanocomposites: MOFs are transition metal centres or clusters combined with organic ligands, constituting multidimensional structures having tunable pores and hence, completely accessible ultra-high surface area [112]. Their redox-active metal centres, combined with a large accessible surface area, act as a backbone that makes them excellent candidates for energy storage applications [113]. However, poor electrical conductivity, insufficient stability, and limited cycle life are the main disadvantages of these materials alone in electrochemical applications [113]. Developing novel combinations of MOFs having redox active centres with conducting networks like graphene can benefit from superior specific capacitance and enhanced electron transfer properties due to the porous structure and vast number of active sites [114]. This also overcomes the difficulty in establishing organic redox centres in supercapacitor electrodes due to improved electrochemical properties and higher mechanical stability. MOFs integrated with conductive graphene matrices show superior supercapacitor performance compared to traditional metal-based graphene nanocomposites [112].

Graphene-Metal Sulphide Nanocomposites: Metal sulphides are low-cost materials with variable oxidation states, which, on combination with graphene, are known to offer excellent specific capacitance [115]. Compared to metal oxides, metal sulphides have sulphur, capable of exhibiting better performance. However, metal sulphides also fail to simultaneously deliver all the necessary characteristics of an excellent supercapacitor electrode material [116]. As said before, the conducting graphene framework, on hybridizing with this class of compounds, showcases excellent activity. Sulphides like NiS_x, CoS_x, $NiCo_xS_y$, MoS_2, WS_2 etc., are most renowned for their hybrid materials with graphene, with superior supercapacitor performance [117]. Among them, MoS_2 and WS_2, having an additional advantage of a 2D layered structure coherent with that of graphene, are known to have frequent usage in supercapacitor electrodes [116].

Graphene-Carbon Nanocomposites: While electrode materials composed of pure graphene offer impressive specific capacitance, their real-world applicability is hindered by several practical limitations [118]. So, along with their nanocomposites with various pseudocapacitive materials, combinations with other carbon materials are also used to enhance the applicability. Graphene can be combined with carbon materials like CNTs, activated carbons, carbon spheres, carbon dots, bio-materials derived carbons, etc. [17]. Incorporation of carbon-based materials inhibits possible restacking of graphene sheets and improves the energy density of supercapacitors [73]. They can also be built to 3D architectures, thereby showing a tremendous increase in the energy storage capability [115]. Combination of nanoporous carbon with graphene is another possibility that aims maximum pore utilization, enabling fruitful graphene incorporation and maximized active surface area, leading to excellent specific capacitance [17]. Rather than this, various structurally modified graphene is also widely used as supercapacitor electrode materials [17,119].

Graphene-Layered Material Nanocomposites: The synergistic combination of graphene with other 2D materials mitigates the tendency of both components to self-restack, enabling alternative assembly structures that enhance supercapacitor performance by maximizing the individual surface area contributions of each material [54]. MXenes are a novel combination of transition metal carbides and/or nitrides with the general formula $M_{n+1}X_nT_x$, where M represents an early transition metal, X can be C and/or N, and T represents surface functionalities [54,61]. g-C_3N_4, a non-metallic 2D material when exfoliated, exhibits promising supercapacitor applications when combined with graphene [120]. The resulting composite displays pseudocapacitive behaviour attributed to both constituents, and numerous studies have reported on the effectiveness of such combinations [61].

Challenges in Graphene-based Supercapacitors

Even though graphene is a potent material to replace commercially relevant activated carbon-based electrode materials in supercapacitors due to its exceptional properties, there are various challenges which have been briefly mentioned before, which hinders its successful commercialization [22]. Supercapacitors are evolving due to the capability to replace conventional energy storage devices [12]. Specific capacitance is one of the key metrics used to assess the suitability of an electrode material for supercapacitors. However, pristine graphene often falls short of achieving its theoretical capacitance when used as an electrode, which remains a significant practical challenge in its supercapacitor fabrication [121]. It is due to the prominent restacking issues with pristine graphene, which lowers the surface area that hindering the effective double layer formation during charge storage, which limits the attainable specific capacitance and significantly affecting the stability of supercapacitors [9]. Low conductivity, lower ion accessible surface area, and hence low specific capacitance remain as the major limitations of currently available GO-based graphene [122]. Highly precise bottom-up methods are both expensive and low-yielding. Hence, the development of synthetic routes that are green, scalable, cost-effective, and provide highly conducting, high surface area high-quality graphene is another major challenge [123]. The lower conductivity of graphene from graphite oxide assisted methods arises from its broken aromatic π-network due to strong oxidising agents used in the reaction, and hence, future demands graphene synthesis in a more sustainable way to obtain graphene with unaffected conductivity, which in turn acts as a reliable electrode material in supercapacitors [124]. The compromise over conductivity by relying on graphite oxide-assisted methods also causes limited power density and rate capability of supercapacitors made up of those electrodes [118]. Among plenty of graphene-based supercapacitors, restacking and structural defects remain a major challenge [23].

Rather than the highly efficient electrode material, perfect assembly of the device is also essential for a stable performance of a supercapacitor. The increased areal loadings will decrease energy density of the device, and hence developing thin electrodes out of graphene, evading restacking possibilities, has a major impact [118]. Effective composite preparation is also challenging, as the resultant material should exhibit major characteristics of both graphene and the pseudocapacitive component [73]. Along with that, the choice of a proper current collector, suitable electrolyte, separator, etc., affects the performance of the supercapacitor [24]. Graphene also has better chemical and thermal stability than the majority of pseudocapacitive materials, the retention of this peculiarity even in its nanocomposites is also challenging since it demands efficient synthesis [105]. This can directly affect the cycling life of a supercapacitor. Self-discharge due to impurities is another major challenge, where the voltage drop occurs right after charging, and it will greatly affect the stability of supercapacitors [125]. Moreover, most methods used to suppress self-discharge led to a reduction in specific capacitance and deterioration of rate performance [125]. So, achieving advancements to avoid self-discharge affecting other major parameters of supercapacitors remains a substantial hurdle [126]. Green perspectives promote the use of aqueous electrolytes, which cannot offer a voltage more than 1.2 V, instead of toxic solvents that have a wide range [127]. So, supercapacitor devices having a wide potential window, sticking to green aspects, remain as another major challenge. The cycling life and durability of supercapacitors also demand a significant improvement [10].

Strategies for Enhancing Stability and Durability

Cyclic stability of a supercapacitor refers to its ability to retain its initial performance after undergoing numerous charge-discharge cycles or over a defined period of use [10]. This parameter is crucial for evaluating supercapacitors, as they are designed to deliver energy reliably over extended durations, enduring frequent cycling, unlike conventional batteries, which typically have shorter lifespans [128]. Durability refers to a supercapacitor's ability to maintain its stored voltage over an extended period without significant self-discharge, reflecting how well it preserves charge when not in use [127]. It indicates the shelf life of the supercapacitor. Stability and durability are both critical factors that define the overall performance and real-world applicability of a supercapacitor [129]. The stability is related to the energy storage mechanism, and EDLC-based graphene electrodes are known to have excellent cyclic stability [44]. In order to enhance the stability and durability, there are various strategies ranging from designing of electrode material and precise device assembly [10]. And it is difficult to incorporate all these key properties simultaneously in a single device for a stable performance. The rational designing of the electrode by modified graphene structures, like 2D sheets to 3D porous networks is encouraged as it prevents the restacking of graphene sheets [130]. These modifications improve the electrochemical properties of electrode materials, like the porosity, providing an easy diffusion channel for the electrolyte ions [131]. Tailoring various oxygen functionalities over graphene in GO synthetic methods can also enhance the compositing capabilities of graphene-based electrode materials, but at the expense of conductivity [88]. In addition to edge oxygen functionalities, ball mill-assisted synthesis introduced much efficient heteroatom doping to graphene to improve its supercapacitor performance and contribute to excellent stability during practical applications [104]. Graphene-based nanocomposites also help to showcase the excellent cyclic stability of supercapacitors. Among graphene-metal oxide nanocomposites, oxides of Mn in combination with graphene have high stability [128]. The control over graphene microstructure and the change in its composition are also helpful. The choice of electrolyte, fabrication of electrodes, avoiding environmental contaminations, and perfect contact inside a supercapacitor can also result in supercapacitors of improved stability/durability [7]. Graphene microstructure controls fast electron movement, and hence electrochemical process can be stable. Along with this, great concern is given to the fabrication of supercapacitor devices [128]. Asymmetric supercapacitors (ASC) are an area of increased research interest due to the excellent stability of reported systems having a wide voltage window with improved energy and power densities [132]. Combination of supercapacitor electrode materials with batteries known as supercapattery device is another emerging strategy to solve the lower cyclic stability of batteries and low energy density of supercapacitors [36].

Self-discharge is one of the primary challenges to solve since it is affecting the durability of graphene-based supercapacitors [125]. The devices can be made more durable by adapting device assembly that prevents aqueous electrolyte evaporation and impurities, and also by using solid-state electrolytes to terminate self-discharge [126]. Durability can also refer to the ability of flexible or bendable supercapacitors to maintain stable performance after repeated bending or under continuous mechanical deformation, whereas graphene-based electrodes succeed in this regard [28,29]. In addition, the durability of electrode materials based on graphene can be improved by strategizing efficient electrode-electrolyte interaction, even in the flexible systems attainable by accurate fabrication [10].

Conclusion and Future perspectives

Supercapacitors have emerged as a solution to growing energy concerns all over the world due to their great potential for efficient energy storage, quick release, environmental friendliness, exceptional cycle life, and long-term durability, as discussed. The persistent pollution from batteries due to their lower endurance in cycle life opposes their advancement in energy storage. Graphene-based electrodes can be considered as the major component in supercapacitors that offer promising performance along with low pollution. This discussion focused on different graphene-based electrode materials for supercapacitors, various graphene synthesis methods and both of their role in enhancing energy and power densities, the key challenges in achieving highly stable and durable graphene-based supercapacitors, and also the strategies employed to improve their performance. Extensive research is being conducted globally, emphasizing the enhancement of energy and power densities along with retained cyclic stability for supercapacitors. Low-cost graphene produced through mechanochemical exfoliation, with its superior conductivity, larger surface area, abundant edge functional groups, and tunable properties, has the potential to replace GO/rGO-based nanocomposites in supercapacitor applications.

Focusing on the sustainable development of graphene-based electrodes in supercapacitors highlights graphene's potential to address future energy deficiencies. A major advancement has to be attained in terms of lowering the cost of commercial electrode material preparation, which makes the technology more applicable. Along with solving problems regarding self-discharge affecting durability of the device, the fabrication of more versatile and flexible microelectronics can serve as a promising direction for future research. Supercapacitors of high-energy density with preserved excellent power-density are also demanded by the future. The electrophilicity of graphene-based materials can also be utilized for efficient electrode fabrication in the future. Even though there are various promising electrode materials, research has to focus further on commercializing excellent supercapacitors over conventional batteries, and this growth will result in great contributions from devices like supercapattery. Advanced characterization of graphene-based materials can further clarify the mechanism of energy storage in supercapacitors, and hence, more attention can be paid to improving stability and durability during real-world applications.

References

[1] P. Huo, P. Zhao, Y. Wang, B. Liu, G. Yin, M. Dong, A Roadmap for Achieving Sustainable Energy Conversion and Storage: Graphene-Based Composites Used Both as an Electrocatalyst for Oxygen Reduction Reactions and an Electrode Material for a Supercapacitor, Energies 11 (2018) 167. https://doi.org/10.3390/en11010167

[2] P. Lamba, P. Singh, P. Singh, P. Singh, Bharti, A. Kumar, M. Gupta, Y. Kumar, Recent advancements in supercapacitors based on different electrode materials: Classifications, synthesis methods and comparative performance, J. Energy Storage 48 (2022) 103871. https://doi.org/10.1016/j.est.2021.103871

[3]B. Muruganantham, R. Gnanadass, N.P. Padhy, Challenges with renewable energy sources and storage in practical distribution systems, Renew. Sustain. Energy Rev. 73 (2017) 125–134. https://doi.org/10.1016/j.rser.2017.01.089

[4]J. Yan, Q. Wang, T. Wei, Z. Fan, Recent Advances in Design and Fabrication of Electrochemical Supercapacitors with High Energy Densities, Adv. Energy Mater. 4 (2014) 1300816. https://doi.org/10.1002/aenm.201300816

[5]J. Libich, J. Máca, J. Vondrák, O. Čech, M. Sedlaříková, Supercapacitors: Properties and applications, J. Energy Storage 17 (2018) 224–227. https://doi.org/10.1016/j.est.2018.03.012

[6] O.S. Adedoja, E.R. Sadiku, Y. Hamam, An Overview of the Emerging Technologies and Composite Materials for Supercapacitors in Energy Storage Applications, Polymers 15 (2023) 2272. https://doi.org/10.3390/polym15102272

[7]A. Shuja, H.R. Khan, I. Murtaza, S. Ashraf, Y. Abid, F. Farid, F. Sajid, Supercapacitors for energy storage applications: Materials, devices and future directions: A comprehensive review, J. Alloys Compd. 1009 (2024) 176924. https://doi.org/10.1016/j.jallcom.2024.176924

[8]C.V.V. Muralee Gopi, S. Alzahmi, V. Narayanaswamy, R. Vinodh, B. Issa, I.M. Obaidat, Supercapacitors: A promising solution for sustainable energy storage and diverse applications, J. Energy Storage 114 (2025) 115729. https://doi.org/10.1016/j.est.2025.115729

[9]S.K. Tiwari, A.K. Thakur, A.D. Adhikari, Y. Zhu, N. Wang, Current Research of Graphene-Based Nanocomposites and Their Application for Supercapacitors, Nanomaterials 10 (2020) 2046. https://doi.org/10.3390/nano10102046

[10] Q. Wu, T. He, Y. Zhang, J. Zhang, Z. Wang, Y. Liu, L. Zhao, Y. Wu, F. Ran, Cyclic stability of supercapacitors: materials, energy storage mechanism, test methods, and device, J. Mater. Chem. A 9 (2021) 24094–24147. https://doi.org/10.1039/D1TA06815F

[11] G. Wang, L. Zhang, J. Zhang, A review of electrode materials for electrochemical supercapacitors, Chem Soc Rev 41 (2012) 797–828. https://doi.org/10.1039/C1CS15060J

[12] R.T. Yadlapalli, R.R. Alla, R. Kandipati, A. Kotapati, Super capacitors for energy storage: Progress, applications and challenges, J. Energy Storage 49 (2022) 104194. https://doi.org/10.1016/j.est.2022.104194

[13] Y.B. Tan, J.-M. Lee, Graphene for supercapacitor applications, J. Mater. Chem. A 1 (2013) 14814. https://doi.org/10.1039/c3ta12193c

[14] H. Rashid Khan, A. Latif Ahmad, Supercapacitors: Overcoming current limitations and charting the course for next-generation energy storage, J. Ind. Eng. Chem. 141 (2025) 46–66. https://doi.org/10.1016/j.jiec.2024.07.014

[15] P. Forouzandeh, V. Kumaravel, S.C. Pillai, Electrode Materials for Supercapacitors: A Review of Recent Advances, Catalysts 10 (2020) 969. https://doi.org/10.3390/catal10090969

[16] W. Lv, Z. Li, Y. Deng, Q.-H. Yang, F. Kang, Graphene-based materials for electrochemical energy storage devices: Opportunities and challenges, Energy Storage Mater. 2 (2016) 107–138. https://doi.org/10.1016/j.ensm.2015.10.002

[17] A. Borenstein, O. Hanna, R. Attias, S. Luski, T. Brousse, D. Aurbach, Carbon-based composite materials for supercapacitor electrodes: a review, J. Mater. Chem. A 5 (2017) 12653–12672. https://doi.org/10.1039/C7TA00863E

[18] A. Patel, S.K. Patel, R.S. Singh, R.P. Patel, Review on recent advancements in the role of electrolytes and electrode materials on supercapacitor performances, Discov. Nano 19 (2024) 188. https://doi.org/10.1186/s11671-024-04053-1

[19] A. Abdisattar, M. Yeleuov, C. Daulbayev, K. Askaruly, A. Tolynbekov, A. Taurbekov, N. Prikhodko, Recent advances and challenges of current collectors for supercapacitors, Electrochem. Commun. 142 (2022) 107373. https://doi.org/10.1016/j.elecom.2022.107373

[20] T. Bagarti, A.M. Jayannavar, Storage of Electrical Energy, Resonance 25 (2020) 963–980. https://doi.org/10.1007/s12045-020-1012-0

[21] S.W. Bokhari, A.H. Siddique, P.C. Sherrell, X. Yue, K.M. Karumbaiah, S. Wei, A.V. Ellis, W. Gao, Advances in graphene-based supercapacitor electrodes, Energy Rep. 6 (2020) 2768–2784. https://doi.org/10.1016/j.egyr.2020.10.001

[22] M. Horn, B. Gupta, J. MacLeod, J. Liu, N. Motta, Graphene-based supercapacitor electrodes: Addressing challenges in mechanisms and materials, Curr. Opin. Green Sustain. Chem. 17 (2019) 42–48. https://doi.org/10.1016/j.cogsc.2019.03.004

[23] H.H. Hegazy, J. Khan, N. Shakeel, E.A. Alabdullkarem, M.I. Saleem, H. Alrobei, I.S. Yahia, 2D-based electrode materials for supercapacitors – status, challenges, and prospects, RSC Adv. 14 (2024) 32958–32977. https://doi.org/10.1039/D4RA05473C

[24] Z. Yan, S. Luo, Q. Li, Z.-S. Wu, S. (Frank) Liu, Recent Advances in Flexible Wearable Supercapacitors: Properties, Fabrication, and Applications, Adv. Sci. 11 (2024) 2302172. https://doi.org/10.1002/advs.202302172

[25] J.O. Dennis, M.F. Shukur, O.A. Aldaghri, K.H. Ibnaouf, A.A. Adam, F. Usman, Y.M. Hassan, A. Alsadig, W.L. Danbature, B.A. Abdulkadir, A Review of Current Trends on Polyvinyl Alcohol (PVA)-Based Solid Polymer Electrolytes, Molecules 28 (2023) 1781. https://doi.org/10.3390/molecules28041781

[26] Y. Zhang, H. Mei, Y. Cao, X. Yan, J. Yan, H. Gao, H. Luo, S. Wang, X. Jia, L. Kachalova, J. Yang, S. Xue, C. Zhou, L. Wang, Y. Gui, Recent advances and challenges of electrode materials for flexible supercapacitors, Coord. Chem. Rev. 438 (2021) 213910. https://doi.org/10.1016/j.ccr.2021.213910

[27] M.R. Benzigar, V.D.B.C. Dasireddy, X. Guan, T. Wu, G. Liu, Advances on Emerging Materials for Flexible Supercapacitors: Current Trends and Beyond, Adv. Funct. Mater. 30 (2020) 2002993. https://doi.org/10.1002/adfm.202002993

[28] Flexible supercapacitors based on vertical graphene/carbon fabric with high rate performance, Appl. Surf. Sci. 610 (2023) 155535. https://doi.org/10.1016/j.apsusc.2022.155535

[29] T. Chen, L. Dai, Flexible supercapacitors based on carbon nanomaterials, J. Mater. Chem. A 2 (2014) 10756–10775. https://doi.org/10.1039/C4TA00567H

[30] A. Mishra, N.P. Shetti, S. Basu, K. Raghava Reddy, T.M. Aminabhavi, Carbon Cloth-based Hybrid Materials as Flexible Electrochemical Supercapacitors, ChemElectroChem 6 (2019) 5771–5786. https://doi.org/10.1002/celc.201901122

[31] M. Shaker, A.A. Sadeghi Ghazvini, S. Feng, W. Cao, X. Meng, Q. Ge, R. Riahifar, Improving the Electrochemical Performance of Pouch Cell Electric Double-Layer Capacitors by Integrating Graphene Nanoplates into Activated Carbon, Energy Technol. 10 (2022) 2100735. https://doi.org/10.1002/ente.202100735

[32] X. Zhao, B. Zheng, T. Huang, C. Gao, Graphene-based single fiber supercapacitor with a coaxial structure, Nanoscale 7 (2015) 9399–9404. https://doi.org/10.1039/C5NR01737H

[33] T.B. Naveen, D. Durgalakshmi, A.K. Kunhiraman, S. Balakumar, R. Ajay Rakkesh, Recent advances in graphene-based micro-supercapacitors: Processes and applications, J. Mater. Res. 36 (2021) 4102–4119. https://doi.org/10.1557/s43578-021-00366-4

[34] M.Z. Iqbal, M.M. Faisal, S.R. Ali, Integration of supercapacitors and batteries towards high-performance hybrid energy storage devices, Int. J. Energy Res. 45 (2021) 1449–1479. https://doi.org/10.1002/er.5954

[35] M. Czagany, S. Hompoth, A.K. Keshri, N. Pandit, I. Galambos, Z. Gacsi, P. Baumli, Supercapacitors: An Efficient Way for Energy Storage Application, Materials 17 (2024) 702. https://doi.org/10.3390/ma17030702

[36] M.Z. Iqbal, U. Aziz, Supercapattery: Merging of battery-supercapacitor electrodes for hybrid energy storage devices, J. Energy Storage 46 (2022) 103823. https://doi.org/10.1016/j.est.2021.103823

[37] A. Patra, N. K, J. Rose Jose, S. Sahoo, B. Chakraborty, C. Sekhar Rout, Understanding the charge storage mechanism of supercapacitors: in situ / operando spectroscopic approaches and theoretical investigations, J. Mater. Chem. A 9 (2021) 25852–25891. https://doi.org/10.1039/D1TA07401F

[38] Yu.M. Volfkovich, Electrochemical Supercapacitors (a Review), Russ. J. Electrochem. 57 (2021) 311–347. https://doi.org/10.1134/S1023193521040108

[39] S. Karthikeyan, B. Narenthiran, A. Sivanantham, L.D. Bhatlu, T. Maridurai, Supercapacitor: Evolution and review, Mater. Today Proc. 46 (2021) 3984–3988. https://doi.org/10.1016/j.matpr.2021.02.526

[40] A.G. Olabi, Q. Abbas, A. Al Makky, M.A. Abdelkareem, Supercapacitors as next generation energy storage devices: Properties and applications, Energy 248 (2022) 123617. https://doi.org/10.1016/j.energy.2022.123617

[41] Y. Lu, S. Zhang, J. Yin, C. Bai, J. Zhang, Y. Li, Y. Yang, Z. Ge, M. Zhang, L. Wei, M. Ma, Y. Ma, Y. Chen, Mesoporous activated carbon materials with ultrahigh mesopore volume and effective specific surface area for high performance supercapacitors, Carbon 124 (2017) 64–71. https://doi.org/10.1016/j.carbon.2017.08.044

[42] A.G. Pandolfo, A.F. Hollenkamp, Carbon properties and their role in supercapacitors, J. Power Sources 157 (2006) 11–27. https://doi.org/10.1016/j.jpowsour.2006.02.065

[43] N. Kumar, S.-B. Kim, S.-Y. Lee, S.-J. Park, Recent Advanced Supercapacitor: A Review of Storage Mechanisms, Electrode Materials, Modification, and Perspectives, Nanomaterials 12 (2022) 3708. https://doi.org/10.3390/nano12203708

[44] N.I. Jalal, R.I. Ibrahim, M.K. Oudah, A review on Supercapacitors: types and components, Phys. Conf. Ser. 1973 (2021) 012015. https://doi.org/10.1088/1742-6596/1973/1/012015

[45] D.S. Silvester, R. Jamil, S. Doblinger, Y. Zhang, R. Atkin, H. Li, Electrical Double Layer Structure in Ionic Liquids and Its Importance for Supercapacitor, Battery, Sensing, and Lubrication Applications, J. Phys. Chem. C 125 (2021) 13707–13720. https://doi.org/10.1021/acs.jpcc.1c03253

[46] P. Sharma, T.S. Bhatti, A review on electrochemical double-layer capacitors, Energy Convers. Manag. 51 (2010) 2901–2912. https://doi.org/10.1016/j.enconman.2010.06.031

[47] K.-C. Tsay, L. Zhang, J. Zhang, Effects of electrode layer composition/thickness and electrolyte concentration on both specific capacitance and energy density of supercapacitor, Electrochimica Acta 60 (2012) 428–436. https://doi.org/10.1016/j.electacta.2011.11.087

[48] Bharti, A. Kumar, G. Ahmed, M. Gupta, P. Bocchetta, R. Adalati, R. Chandra, Y. Kumar, Theories and models of supercapacitors with recent advancements: impact and interpretations, Nano Express 2 (2021) 022004. https://doi.org/10.1088/2632-959X/abf8c2

[49] L. Pilon, H. Wang, A. d'Entremont, Recent Advances in Continuum Modeling of Interfacial and Transport Phenomena in Electric Double Layer Capacitors, J. Electrochem. Soc. 162 (2015) A5158. https://doi.org/10.1149/2.0211505jes

[50] M.-S. Wu, K.-H. Lin, One-step Electrophoretic Deposition of Ni-Decorated Activated-Carbon Film as an Electrode Material for Supercapacitors, J. Phys. Chem. C 114 (2010) 6190–6196. https://doi.org/10.1021/jp9109145

[51] E. Frackowiak, Carbon materials for supercapacitor application, Phys. Chem. Chem. Phys. 9 (2007) 1774–1785. https://doi.org/10.1039/B618139M

[52] C. Li, X. Zhang, K. Wang, X. Sun, G. Liu, J. Li, H. Tian, J. Li, Y. Ma, Scalable Self-Propagating High-Temperature Synthesis of Graphene for Supercapacitors with Superior Power Density and Cyclic Stability, Adv. Mater. 29 (2017) 1604690. https://doi.org/10.1002/adma.201604690

[53] M. Girirajan, A.K. Bojarajan, I.N. Pulidindi, K.N. Hui, S. Sangaraju, An insight into the nanoarchitecture of electrode materials on the performance of supercapacitors, Coord. Chem. Rev. 518 (2024) 216080. https://doi.org/10.1016/j.ccr.2024.216080

[54] C. S, M.A.A. Mohd Abdah, V.N. Thakur, M.S. Govinde Gowda, P. Choudhary, J.B. Sriramoju, D. Rangappa, S. Malik, S. Rustagi, M. Khalid, Progress and Prospects of MXene-Based Hybrid Composites for Next-Generation Energy Technology, J. Electrochem. Soc. 170 (2023) 120530. https://doi.org/10.1149/1945-7111/ad0c64

[55] Y. Wang, L. Zhang, H. Hou, W. Xu, G. Duan, S. He, K. Liu, S. Jiang, Recent progress in carbon-based materials for supercapacitor electrodes: a review, J. Mater. Sci. 56 (2021) 173–200. https://doi.org/10.1007/s10853-020-05157-6

[56] S. Sahoo, R. Kumar, E. Joanni, R.K. Singh, J.-J. Shim, Advances in pseudocapacitive and battery-like electrode materials for high performance supercapacitors, J. Mater. Chem. A 10 (2022) 13190–13240. https://doi.org/10.1039/D2TA02357A

[57] X. Zhu, Recent advances of transition metal oxides and chalcogenides in pseudocapacitors and hybrid capacitors: A review of structures, synthetic strategies, and mechanism studies, J. Energy Storage 49 (2022) 104148. https://doi.org/10.1016/j.est.2022.104148

[58] P. Bhojane, Recent advances and fundamentals of Pseudocapacitors: Materials, mechanism, and its understanding, J. Energy Storage 45 (2022) 103654. https://doi.org/10.1016/j.est.2021.103654

[59] S. Mahala, K. Khosravinia, A. Kiani, Unwanted degradation in pseudocapacitors: Challenges and opportunities, J. Energy Storage 67 (2023) 107558. https://doi.org/10.1016/j.est.2023.107558

[60] T. Schoetz, L.W. Gordon, S. Ivanov, A. Bund, D. Mandler, R.J. Messinger, Disentangling faradaic, pseudocapacitive, and capacitive charge storage: A tutorial for the characterization of batteries, supercapacitors, and hybrid systems, Electrochimica Acta 412 (2022) 140072. https://doi.org/10.1016/j.electacta.2022.140072

[61] S.J. Panchu, K. Raju, H.C. Swart, Emerging Two–Dimensional Intercalation Pseudocapacitive Electrodes for Supercapacitors, ChemElectroChem 11 (2024) e202300810. https://doi.org/10.1002/celc.202300810

[62] H.W. Park, K.C. Roh, Recent advances in and perspectives on pseudocapacitive materials for Supercapacitors–A review, J. Power Sources 557 (2023) 232558. https://doi.org/10.1016/j.jpowsour.2022.232558

[63] W.H. Low, P.S. Khiew, S.S. Lim, C.W. Siong, E.R. Ezeigwe, Recent development of mixed transition metal oxide and graphene/mixed transition metal oxide based hybrid nanostructures for advanced supercapacitors, J. Alloys Compd. 775 (2019) 1324–1356. https://doi.org/10.1016/j.jallcom.2018.10.102

[64] P. Navalpotro, M. Anderson, R. Marcilla, J. Palma, Insights into the energy storage mechanism of hybrid supercapacitors with redox electrolytes by Electrochemical Impedance Spectroscopy, Electrochimica Acta 263 (2018) 110–117. https://doi.org/10.1016/j.electacta.2017.12.167

[65] D. P. Chatterjee, A. K. Nandi, A review on the recent advances in hybrid supercapacitors, J. Mater. Chem. A 9 (2021) 15880–15918. https://doi.org/10.1039/D1TA02505H

[66] A. Afif, S.M. Rahman, A. Tasfiah Azad, J. Zaini, M.A. Islan, A.K. Azad, Advanced materials and technologies for hybrid supercapacitors for energy storage – A review, J. Energy Storage 25 (2019) 100852. https://doi.org/10.1016/j.est.2019.100852

[67] A. Muzaffar, M.B. Ahamed, K. Deshmukh, J. Thirumalai, A review on recent advances in hybrid supercapacitors: Design, fabrication and applications, Renew. Sustain. Energy Rev. 101 (2019) 123–145. https://doi.org/10.1016/j.rser.2018.10.026

[68] Y. Shao, M.F. El-Kady, J. Sun, Y. Li, Q. Zhang, M. Zhu, H. Wang, B. Dunn, R.B. Kaner, Design and Mechanisms of Asymmetric Supercapacitors, Chem. Rev. 118 (2018) 9233–9280. https://doi.org/10.1021/acs.chemrev.8b00252

[69] Q. Ke, J. Wang, Graphene-based materials for supercapacitor electrodes – A review, J. Materiomics 2 (2016) 37–54. https://doi.org/10.1016/j.jmat.2016.01.001

[70] L.L. Zhang, R. Zhou, X.S. Zhao, Graphene-based materials as supercapacitor electrodes, J. Mater. Chem. 20 (2010) 5983. https://doi.org/10.1039/c000417k

[71] A.H. Castro Neto, F. Guinea, N.M.R. Peres, K.S. Novoselov, A.K. Geim, The electronic properties of graphene, Rev. Mod. Phys. 81 (2009) 109–162. https://doi.org/10.1103/RevModPhys.81.109

[72] S.K. Kandasamy, K. Kandasamy, Recent Advances in Electrochemical Performances of Graphene Composite (Graphene-Polyaniline/Polypyrrole/Activated Carbon/Carbon Nanotube) Electrode Materials for Supercapacitor: A Review, J. Inorg. Organomet. Polym. Mater. 28 (2018) 559–584. https://doi.org/10.1007/s10904-018-0779-x

[73] N. Mahmood, C. Zhang, H. Yin, Y. Hou, Graphene-based nanocomposites for energy storage and conversion in lithium batteries, supercapacitors and fuel cells, J Mater Chem A 2 (2014) 15–32. https://doi.org/10.1039/C3TA13033A

[74] K.E. Whitener, P.E. Sheehan, Graphene synthesis, Diam. Relat. Mater. 46 (2014) 25–34. https://doi.org/10.1016/j.diamond.2014.04.006

[75] A. Armano, S. Agnello, Two-Dimensional Carbon: A Review of Synthesis Methods, and Electronic, Optical, and Vibrational Properties of Single-Layer Graphene, C 5 (2019) 67. https://doi.org/10.3390/c5040067

[76] F. Bonaccorso, A. Lombardo, T. Hasan, Z. Sun, L. Colombo, A.C. Ferrari, Production and processing of graphene and 2D crystals, Mater. Today 15 (2012) 564–589. https://doi.org/10.1016/S1369-7021(13)70014-2

[77] K. Parvez, Z.-S. Wu, R. Li, X. Liu, R. Graf, X. Feng, K. Müllen, Exfoliation of Graphite into Graphene in Aqueous Solutions of Inorganic Salts, J. Am. Chem. Soc. 136 (2014) 6083–6091. https://doi.org/10.1021/ja5017156

[78] H. Yu, B. Zhang, C. Bulin, R. Li, R. Xing, High-efficient Synthesis of Graphene Oxide Based on Improved Hummers Method, Sci. Rep. 6 (2016) 36143. https://doi.org/10.1038/srep36143

[79] W. Wang, Y. Hou, D. Martinez, D. Kurniawan, W.-H. Chiang, P. Bartolo, Carbon Nanomaterials for Electro-Active Structures: A Review, Polymers 12 (2020) 2946. https://doi.org/10.3390/polym12122946

[80] M.G. Sumdani, M.R. Islam, A.N.A. Yahaya, S.I. Safie, Recent advances of the graphite exfoliation processes and structural modification of graphene: a review, J. Nanoparticle Res. 23 (2021) 253. https://doi.org/10.1007/s11051-021-05371-6

[81] R. Ye, J.M. Tour, Graphene at Fifteen, ACS Nano 13 (2019) 10872–10878. https://doi.org/10.1021/acsnano.9b06778

[82] R. C. Sinclair, J. L. Suter, P. V. Coveney, Micromechanical exfoliation of graphene on the atomistic scale, Phys. Chem. Chem. Phys. 21 (2019) 5716–5722. https://doi.org/10.1039/C8CP07796G

[83] M. Yi, Z. Shen, A review on mechanical exfoliation for the scalable production of graphene, J. Mater. Chem. A 3 (2015) 11700–11715. https://doi.org/10.1039/C5TA00252D

[84] B. Jayasena, S. Subbiah, A novel mechanical cleavage method for synthesizing few-layer graphenes, Nanoscale Res. Lett. 6 (2011) 95. https://doi.org/10.1186/1556-276X-6-95

[85] A.F. Betancur, N. Ornelas-Soto, A.M. Garay-Tapia, F.R. Pérez, Á. Salazar, A.G. García, A general strategy for direct synthesis of reduced graphene oxide by chemical exfoliation of graphite, Mater. Chem. Phys. 218 (2018) 51–61. https://doi.org/10.1016/j.matchemphys.2018.07.019

[86] N. Cao, Y. Zhang, Study of Reduced Graphene Oxide Preparation by Hummers' Method and Related Characterization, J. Nanomater. 2015 (2015) 168125. https://doi.org/10.1155/2015/168125

[87] J. Chen, B. Yao, C. Li, G. Shi, An improved Hummers method for eco-friendly synthesis of graphene oxide, Carbon 64 (2013) 225–229. https://doi.org/10.1016/j.carbon.2013.07.055

[88] J. Chen, Y. Li, L. Huang, C. Li, G. Shi, High-yield preparation of graphene oxide from small graphite flakes via an improved Hummers method with a simple purification process, Carbon 81 (2015) 826–834. https://doi.org/10.1016/j.carbon.2014.10.033

[89] S. Balasubramanyan, S. Sasidharan, R. Poovathinthodiyil, R. M. Ramakrishnan, B. N. Narayanan, Sucrose-mediated mechanical exfoliation of graphite: a green method for the large scale production of graphene and its application in catalytic reduction of 4-nitrophenol, New J. Chem. 41 (2017) 11969–11978. https://doi.org/10.1039/C7NJ01900A

[90] P. Yu, S.E. Lowe, G.P. Simon, Y.L. Zhong, Electrochemical exfoliation of graphite and production of functional graphene, Curr. Opin. Colloid Interface Sci. 20 (2015) 329–338. https://doi.org/10.1016/j.cocis.2015.10.007

[91] F. Liu, C. Wang, X. Sui, M.A. Riaz, M. Xu, L. Wei, Y. Chen, Synthesis of graphene materials by electrochemical exfoliation: Recent progress and future potential, Carbon Energy 1 (2019) 173–199. https://doi.org/10.1002/cey2.14

[92] A. M. Abdelkader, A. J. Cooper, R.A. W. Dryfe, I. A. Kinloch, How to get between the sheets: a review of recent works on the electrochemical exfoliation of graphene materials from bulk graphite, Nanoscale 7 (2015) 6944–6956. https://doi.org/10.1039/C4NR06942K

[93] W. Du, X. Jiang, L. Zhu, From graphite to graphene: direct liquid-phase exfoliation of graphite to produce single- and few-layered pristine graphene, J. Mater. Chem. A 1 (2013) 10592. https://doi.org/10.1039/c3ta12212c

[94] Y. Hernandez, V. Nicolosi, M. Lotya, F.M. Blighe, Z. Sun, S. De, I.T. McGovern, B. Holland, M. Byrne, Y.K. Gun'Ko, J.J. Boland, P. Niraj, G. Duesberg, S. Krishnamurthy, R. Goodhue, J. Hutchison, V. Scardaci, A.C. Ferrari, J.N. Coleman, High-yield production of graphene by liquid-phase exfoliation of graphite, Nat. Nanotechnol. 3 (2008) 563–568. https://doi.org/10.1038/nnano.2008.215

[95] A. Ciesielski, P. Samorì, Graphene via sonication assisted liquid-phase exfoliation, Chem. Soc. Rev. 43 (2014) 381–398. https://doi.org/10.1039/C3CS60217F

[96] Y. Xu, H. Cao, Y. Xue, B. Li, W. Cai, Liquid-Phase Exfoliation of Graphene: An Overview on Exfoliation Media, Techniques, and Challenges, Nanomaterials 8 (2018) 942. https://doi.org/10.3390/nano8110942

[97] D. Parviz, F. Irin, S.A. Shah, S. Das, C.B. Sweeney, M.J. Green, Challenges in Liquid-Phase Exfoliation, Processing, and Assembly of Pristine Graphene, Adv. Mater. 28 (2016) 8796–8818. https://doi.org/10.1002/adma.201601889

[98] Chandni. A p, S. Vattapparambil Chandran, B.N. Narayanan, An environmentally sustainable ultrasonic-assisted exfoliation approach to graphene and its nanocompositing with polyaniline for supercapacitor applications, Ultrasonics 145 (2025) 107482. https://doi.org/10.1016/j.ultras.2024.107482

[99] Md.S.A. Bhuyan, Md.N. Uddin, Md.M. Islam, F.A. Bipasha, S.S. Hossain, Synthesis of graphene, Int. Nano Lett. 6 (2016) 65–83. https://doi.org/10.1007/s40089-015-0176-1

[100] K.S. Subrahmanyam, L.S. Panchakarla, A. Govindaraj, C.N.R. Rao, Simple Method of Preparing Graphene Flakes by an Arc-Discharge Method, J. Phys. Chem. C 113 (2009) 4257–4259. https://doi.org/10.1021/jp900791y

[101] Y. Zhang, L. Zhang, C. Zhou, Review of Chemical Vapor Deposition of Graphene and Related Applications, Acc. Chem. Res. 46 (2013) 2329–2339. https://doi.org/10.1021/ar300203n

[102] X.Z. Yu, C.G. Hwang, C.M. Jozwiak, A. Köhl, A.K. Schmid, A. Lanzara, New synthesis method for the growth of epitaxial graphene, J. Electron Spectrosc. Relat. Phenom. 184 (2011) 100–106. https://doi.org/10.1016/j.elspec.2010.12.034

[103] H.-F. Yen, Y.-Y. Horng, M.-S. Hu, W.-H. Yang, J.-R. Wen, A. Ganguly, Y. Tai, K.-H. Chen, L.-C. Chen, Vertically aligned epitaxial graphene nanowalls with dominated nitrogen doping for superior supercapacitors, Carbon 82 (2015) 124–134. https://doi.org/10.1016/j.carbon.2014.10.042

[104] X. Fan, D.W. Chang, X. Chen, J.-B. Baek, L. Dai, Functionalized graphene nanoplatelets from ball milling for energy applications, Curr. Opin. Chem. Eng. 11 (2016) 52–58. https://doi.org/10.1016/j.coche.2016.01.003

[105] S.S. Shams, R. Zhang, J. Zhu, Graphene synthesis: a Review, Mater. Sci.-Pol. 33 (2015) 566–578. https://doi.org/10.1515/msp-2015-0079

[106] W. Zhao, M. Fang, F. Wu, H. Wu, L. Wang, G. Chen, Preparation of graphene by exfoliation of graphite using wet ball milling, J. Mater. Chem. 20 (2010) 5817–5819. https://doi.org/10.1039/C0JM01354D

[107] A.E.D. Mahmoud, A. Stolle, M. Stelter, Sustainable Synthesis of High-Surface-Area Graphite Oxide via Dry Ball Milling, ACS Sustain. Chem. Eng. 6 (2018) 6358–6369. https://doi.org/10.1021/acssuschemeng.8b00147

[108] R. Lakra, R. Kumar, P.K. Sahoo, D. Thatoi, A. Soam, A mini-review: Graphene based composites for supercapacitor application, Inorg. Chem. Commun. 133 (2021) 108929. https://doi.org/10.1016/j.inoche.2021.108929

[109] D. Nandi, V.B. Mohan, A.K. Bhowmick, D. Bhattacharyya, Metal/metal oxide decorated graphene synthesis and application as supercapacitor: a review, J. Mater. Sci. 55 (2020) 6375–6400. https://doi.org/10.1007/s10853-020-04475-z

[110] F. Ahmad, M. Zahid, H. Jamil, M.A. Khan, S. Atiq, M. Bibi, K. Shahbaz, M. Adnan, M. Danish, F. Rasheed, H. Tahseen, M.J. Shabbir, M. Bilal, A. Samreen, Advances in graphene-based electrode materials for high-performance supercapacitors: A review, J. Energy Storage 72 (2023) 108731. https://doi.org/10.1016/j.est.2023.108731

[111] X. Sun, C. Huang, L. Wang, L. Liang, Y. Cheng, W. Fei, Y. Li, Recent Progress in Graphene/Polymer Nanocomposites, Adv. Mater. 33 (2021) 2001105. https://doi.org/10.1002/adma.202001105

[112] D.K. Singha, R.I. Mohanty, P. Bhanja, B.K. Jena, Metal–organic framework and graphene composites: advanced materials for electrochemical supercapacitor applications, Mater. Adv. 4 (2023) 4679–4706. https://doi.org/10.1039/D3MA00523B

[113] L.G. Beka, X. Bu, X. Li, X. Wang, C. Han, W. Liu, A 2D metal–organic framework/reduced graphene oxide heterostructure for supercapacitor application, RSC Adv. 9 (2019) 36123–36135. https://doi.org/10.1039/C9RA07061C

[114] P. Srimuk, S. Luanwuthi, A. Krittayavathananon, M. Sawangphruk, Solid-type supercapacitor of reduced graphene oxide-metal organic framework composite coated on carbon fiber paper, Electrochimica Acta 157 (2015) 69–77. https://doi.org/10.1016/j.electacta.2015.01.082

[115] J. Ran, Y. Liu, H. Fong, H. Shi, Q. Ma, A review on graphene based electrode materials for supercapacitor, J. Ind. Eng. Chem. 137 (2024) 106–121. https://doi.org/10.1016/j.jiec.2024.03.043

[116] P. Geng, S. Zheng, H. Tang, R. Zhu, L. Zhang, S. Cao, H. Xue, H. Pang, Transition Metal Sulfides Based on Graphene for Electrochemical Energy Storage, Adv. Energy Mater. 8 (2018) 1703259. https://doi.org/10.1002/aenm.201703259

[117] F. Yu, Z. Chang, X. Yuan, F. Wang, Y. Zhu, L. Fu, Y. Chen, H. Wang, Y. Wu, W. Li, Ultrathin $NiCo_2S_4$@graphene with a core–shell structure as a high performance positive electrode for hybrid supercapacitors, J. Mater. Chem. A 6 (2018) 5856–5861. https://doi.org/10.1039/C8TA00835C

[118] H. Zhang, D. Yang, A. Lau, T. Ma, H. Lin, B. Jia, Hybridized Graphene for Supercapacitors: Beyond the Limitation of Pure Graphene, Small 17 (2021) 2007311. https://doi.org/10.1002/smll.202007311

[119] J. Huang, J. Wang, C. Wang, H. Zhang, C. Lu, J. Wang, Hierarchical Porous Graphene Carbon-Based Supercapacitors, Chem. Mater. 27 (2015) 2107–2113. https://doi.org/10.1021/cm504618r

[120] A.R. Sonkawade, S.S. Mahajan, A.R. Shelake, S.A. Ahir, M.R. Waikar, S.S. Sutar, R.G. Sonkawade, T.D. Dongale, The g-C_3N_4/rGO composite for high-performance supercapacitor: Synthesis, characterizations, and time series modeling and predictions, Int. J. Hydrog. Energy 87 (2024) 1416–1426. https://doi.org/10.1016/j.ijhydene.2024.09.129

[121] S. Zheng, Z.-S. Wu, S. Wang, H. Xiao, F. Zhou, C. Sun, X. Bao, H.-M. Cheng, Graphene-based materials for high-voltage and high-energy asymmetric supercapacitors, Energy Storage Mater. 6 (2017) 70–97. https://doi.org/10.1016/j.ensm.2016.10.003

[122] R.R. Salunkhe, Y.-H. Lee, K.-H. Chang, J.-M. Li, P. Simon, J. Tang, N.L. Torad, C.-C. Hu, Y. Yamauchi, Nanoarchitectured Graphene-Based Supercapacitors for Next-Generation Energy-Storage Applications, Chem. – Eur. J. 20 (2014) 13838–13852. https://doi.org/10.1002/chem.201403649

[123] X. Shi, S. Zheng, Z.-S. Wu, X. Bao, Recent advances of graphene-based materials for high-performance and new-concept supercapacitors, J. Energy Chem. 27 (2018) 25–42. https://doi.org/10.1016/j.jechem.2017.09.034

[124] S. Rao, J. Upadhyay, K. Polychronopoulou, R. Umer, R. Das, Reduced Graphene Oxide: Effect of Reduction on Electrical Conductivity, J. Compos. Sci. 2 (2018) 25. https://doi.org/10.3390/jcs2020025

[125] W. Li, W. Yang, M. Wu, M. Zhao, X. Lu, Polydopamine-coated graphene for supercapacitors with improved electrochemical performances and reduced self-discharge, Electrochimica Acta 426 (2022) 140776. https://doi.org/10.1016/j.electacta.2022.140776

[126] R. Yuan, Y. Dong, R. Hou, S. Zhang, H. Song, Review—Influencing Factors and Suppressing Strategies of the Self-Discharge for Carbon Electrode Materials in Supercapacitors, J. Electrochem. Soc. 169 (2022) 030504. https://doi.org/10.1149/1945-7111/ac56a1

[127] O. Okhay, A. Tkach, P. Staiti, F. Lufrano, Long term durability of solid-state supercapacitor based on reduced graphene oxide aerogel and carbon nanotubes composite electrodes, Electrochimica Acta 353 (2020) 136540. https://doi.org/10.1016/j.electacta.2020.136540

[128] A.W. Anwar, A. Majeed, N. Iqbal, W. Ullah, A. Shuaib, U. Ilyas, F. Bibi, H.M. Rafique, Specific Capacitance and Cyclic Stability of Graphene Based Metal/Metal Oxide Nanocomposites: A Review, J. Mater. Sci. Technol. 31 (2015) 699–707. https://doi.org/10.1016/j.jmst.2014.12.012

[129] S.-K. Kim, H.J. Kim, J.-C. Lee, P.V. Braun, H.S. Park, Extremely Durable, Flexible Supercapacitors with Greatly Improved Performance at High Temperatures, ACS Nano 9 (2015) 8569–8577. https://doi.org/10.1021/acsnano.5b03732

[130] Z. Yang, S. Chabi, Y. Xia, Y. Zhu, Preparation of 3D graphene-based architectures and their applications in supercapacitors, Prog. Nat. Sci. Mater. Int. 25 (2015) 554–562. https://doi.org/10.1016/j.pnsc.2015.11.010

[131] X. Cao, Y. Shi, W. Shi, G. Lu, X. Huang, Q. Yan, Q. Zhang, H. Zhang, Preparation of Novel 3D Graphene Networks for Supercapacitor Applications, Small 7 (2011) 3163–3168. https://doi.org/10.1002/smll.201100990

[132] H. Wang, Y. Liang, T. Mirfakhrai, Z. Chen, H.S. Casalongue, H. Dai, Advanced asymmetrical supercapacitors based on graphene hybrid materials, Nano Res. 4 (2011) 729–736. https://doi.org/10.1007/s12274-011-0129-6

Electrocatalysts and Advanced Materials for Sustainable Energy Storage   Materials Research Forum LLC
Materials Research Foundations 182 (2025) 160-173   https://doi.org/10.21741/9781644903797-11

Chapter 11

Biomass to Energy: A Sustainable Pathway for Energy Storage

M. PRIYADARSHINI[1,a*], P. VIJAYALAKSHMI[1,b]

[1]Tamil Nadu Pollution Control Board, District Environmental Laboratory, Maraimalai Nagar, Chengalpattu – 603203, India

[a]mpiryadarshinitnpcb@gmail.com, [b]vijayalakshmitnpcb@gmail.com

Abstract

In this chapter, an overview of materials derived from biomass as an alternative to environmentally friendly energy storage and conversion devices are presented. The classification and resources of biomass suitable for energy storage and conversion devices are also discussed. Additionally, the detailed explanation of various conversion techniques adopted to obtain materials from biomass resources are explained in detail. It also provides a detailed explanation of different biomass-derived materials and their applications in supercapacitors, metal-ion batteries, redox-flow batteries, and fuel cells. Furthermore, the advantages and disadvantages of biomass-derived materials are compared with those of commercial materials.

Keywords

Biomass, Electrochemical Energy Storage, Supercapacitors, Li-ion Batteries, Redox-flow Batteries, Fuel Cell

Contents

Introduction

The importance of energy cannot be overstated, as energy is a vital part of today's world, starting from the tiniest daily task to the largest technological achievements. Energy is the backbone of development and progress, starting from running our homes and industries to supporting transportation, healthcare, communication, and agriculture. In today's digital age, where technology, data centers, and smart devices are central to our daily routines, reliable energy sources are more important than ever. Due to a paradigm shift in climate change, the world is shifting toward a clean and sustainable environment. Mainly focusing on renewable energy such as solar, wind, tidal, hydropower, to ensure a greener future. These energy resources not only help in a sustainable environment but also contribute to long-term climate resilience and technological innovation [1-3].

The rapid shift to renewable energy systems the world over and the near monopoly of electrified technologies has made the need for improved energy storage and conversion devices with high efficiency, long-term stability, and environmental friendliness even greater. Even though conventional electrochemical materials are better in terms of electrical conductivity, scalability, and stable chemical structure, it is often expensive, lack sustainability and environment friendliness. Biomass derived materials gained considerable attention in the upcoming years due to their copious availability, low cost, structural stability, innate functional groups & other intricate properties ideal for use in electrochemical processes [4-5].

The agricultural wastes, municipal solid waste, forestry residue, livestock wastes, industrial wastes, and other wastes are versatile and renewable precursors for generating a wide range of carbonaceous and polymeric materials. When processed through different conversion methods, these materials can be tailored into highly porous, heteroatom doped, structurally arranged, and thermally stable. The well-structured materials are suitable for electrode, electrolytes, binders and separators for the electrochemical systems [6-7].

In this chapter, we discuss the different types of conversion techniques adopted to convert biomass into other valuable products to utilize in all electrochemical applications. Their unique properties, such as chemical, structural, and thermal, arise individually and differ for the type of conversion process adopted. Further, discussions are extended in the integration of biomass derived materials into diverse energy systems, including metal-ion batteries (e.g, Li-ion, Nai-ion batteries), supercapacitors, redox-flow batteries, and fuel cells. Each section highlights the importance of the role of biomass materials in electrochemical energy storage and conversion devices & possible ways to improve the energy & power density, cycling stability, ionic & electrical conductivity and scalable synthesis.

Apart from these positive attributes, this chapter also examines the challenges and limitations - inconsistent performance, reduced conductivity, limited scalability- encountered by the biomass derived materials. In conclusion, the chapter explores the emerging trends, the inclusion of recent AI & machine learning technologies wisely for material selection, new synthesis methodologies and future research directions. These also aim at overcoming the barriers and paving the way to large-scale utilization of biomass-materials as solutions to next-generation clean energy technologies.

Biomass Resources & Classification of biomass

Biomass is biodegradable organic matter derived from plants, animals, and other living microorganisms [8], and can be used as a valuable resource for generating renewable energy. Also, based on generations, the biomass resources can be classified as primary resources [9] – produced by photosynthesis and taken directly from the land the based on the nature and origin, biomass can be classified into the following categories:

> ➤ Plant-derived Biomass
> ➤ Animal- derived Biomass
> ➤ Micro-organism derived Biomass

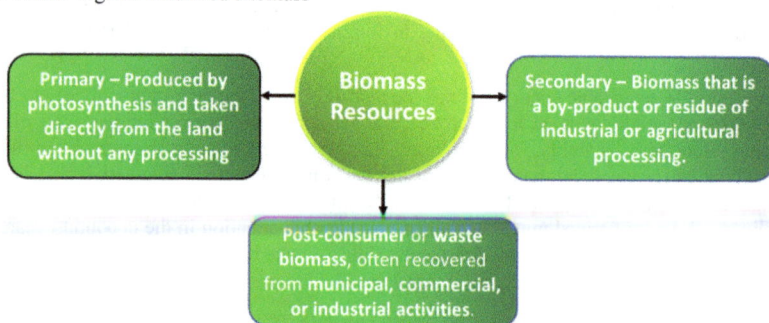

Figure 1. Classification of Biomass resources

Conversion Techniques

The conversion of biomass into valuable products is attained through various techniques which are broadly categorized into thermochemical, biochemical and physiochemical methods [9]. By disintegration of the complex structure of biomass into simpler molecules, these methods convert carbon rich biomass into carbon-based materials, bioenergy products, building block chemicals, materials & composites. For example, rice-husk-derived biomass contains a moisture content of approximately 8-12 %, with a carbon content of 35-50 %, and thermal stability of up to 300 °C [10]. Although it has various properties, the moisture and carbon content significantly affect the working process of energy storage devices. Hence, suitable conversion techniques should be adopted to obtain the best electrode materials for energy storage devices. In addition, the conversion process is classified into various types based on the nature and origin of the biomass as follows:

Thermochemical Conversion

Thermochemical method is the process of converting biomass into value-added products by heat and chemical reactions in the absence/ limited use of oxygen. In higher temperatures of 300 – 1200 °C, the complex structure of biomass are breaks down into simple molecules with the aid of a catalyst. Mainly, this method is preferable for biomass such as wood waste, municipal waste, and agricultural waste [11]. Based on the operating temperature, presence of oxygen, and type of products, thermochemical methods are further classified into combustion, pyrolysis, and hydrothermal conversion processes. The operating temperature, oxygen presence, and product yield during combustion and pyrolysis are listed in Table 1 [4-5, 9, 11].

Table 1. Key Process Conditions for Thermochemical Conversion of Biomass into Electrode Materials

Sl. No	Process	Temperature Range (°C)	Atmosphere	Main Products
1.	Combustion	800 – 1000	Oxygen (Air)	Heat, Carbon dioxide, Water
2.	Pyrolysis	300 – 600	No Oxygen (inert)	Biochar
3.	Hydrothermal Conversion	100 – 250	No Oxygen (water medium)	Biocrude

Biochemical Conversion

Biochemical conversion is the process of disintegration of biomass into valuable products by means of microbial agents such as bacteria, enzymes and other micro-organisms. This method is suitable for converting wet and biodegradable biomass resources such as food waste and sludge [12]. The process is further sub-classified into different types based on the biological actions (i.e., enzyme, type of bacteria used) and presence of oxygen, such as anaerobic digestion, fermentation, composting, and enzymatic hydrolysis. The conditions to be maintained, main products, and presence of oxygen for the biochemical process are listed in Table 2 [4-5, 12].

Table 2. Key Process Conditions for Biochemical Conversion of Biomass into Electrode Materials

Sl. No	Process	Conditions	Products	Oxygen Presence
1.	Anaerobic Digestion	Anaerobic	Biogas	Absence of oxygen
2.	Fermentation	Microbes	Bioethanol	Optional
3.	Composting	Aerobic	Compost	Presence of oxygen
4.	Enzymatic Hydrolysis	Enzymes	Sugars	Absence of oxygen

Physicochemical Conversion:

Physiochemical conversion involves the use of physical and chemical methodologies, often combined or individually, to transform biomass into valuable products. This process involves the use of optimum conditions, such as a special catalyst at ambient temperature, without the need for any microorganisms [13]. The operating temperature, oxygen presence, and product yield during combustion and pyrolysis are listed in Table 3 [4-5, 13].

Table 3. Key Process Conditions for Physiochemical Conversion of Biomass into Electrode Materials

Sl. No	Process	Conditions	Purpose	Product
1.	Carbonization	300 – 600 °C, Inert atmosphere	Convert biomass to carbon	Biochar
2.	Chemical treatment	600 – 900 °C, Chemical activators	Porous network & more surface area	Activated Carbon
3.	Heteroatom incorporation	600–800 °C, Inert atmosphere	Functional group enrichment	N/S/P - doped carbon
4.	Structural Optimization	600–800 °C, inert atmosphere, suitable templates	Construct hierarchical porous networks	Mesoporous or Microporous Carbon

Structural & Physiochemical Properties

Biomass resources from agriculture, animals, and microorganisms have various properties, such as physical, chemical, structural, thermal, and biochemical properties. As discussed above, when biomass is derived from various conversion techniques, its structural and physicochemical properties differ. When biomass resources are converted using different conversion techniques, the intricate properties are enhanced, resulting in carbonaceous materials that are useful as electrode materials for energy storage devices. Once the suitable properties of the biomass-derived materials are identified, the materials can be used for either of the energy storage devices [14].

When utilizing biomass-derived materials for energy storage applications, the derived materials should exhibit properties such as high porosity, high surface area, uniform surface texture, high thermal stability, and conductivity. When converting biomass materials, it contain high carbon content owing to the presence of carbohydrates. These biomass-derived carbon materials have an orderly structure with all the properties, making them particularly suitable for energy storage applications. The presence of carbon, hydrogen, and oxygen in biomass forms complex polymeric structures. When these polymeric structures undergo various processes, the as-derived biomass materials become carbonaceous materials rich in carbon [15-16]. During the thermochemical process, the presence of other elements carbon, are removed as volatile gases such as CO_2, CH_4, and water vapor. The as-derived materials, which refer to biochar and activated carbon, can be directly utilized as electrode materials for energy storage applications. Biomass is mainly composed of lignin, cellulose, and hemicellulose polymeric structures. Wherein, the polymeric cellulose consists of linear chains of β-1,4-linked D-glucose units arranged in tightly packed fibrils and is hydrophobic in nature due to the tight hydrogen bonding. Hemicellulose is a branched heteropolymer of various sugars that surround cellulose, providing flexibility to the structure. 15–30% of biomass contains lignin, a complex aromatic polymer that functions as a binder and resists microbial breakdown and hydrophobicity [17]. Together with cellulose and hemicellulose, it forms a cross-link that fortifies and shields. In addition, biomass contains trace levels of inorganic minerals (also known as ash content) and extractives (such as oils and waxes).

Porosity: When lignocellulosic-rich biomass is converted into carbonaceous materials, the porosity is enhanced, which mainly consists of meso-, micro-, and macro–porous structures. During the conversion of hemicellulose at lower temperatures of $200 - 300\ °C$, it decomposes and releases volatile gases from the solid matrix, leaving micro- and mesopores. Due to the lack of crystalline order of hemicellulose, the unequal and expansive porosity is formed, which allows creation of voids in the carbon structure. These pores might enlarge further and acts a nucleation sites, during physical and chemical activation processes. Hence, the carbon materials obtained from biomass rich in hemicellulose show unique properties such as higher surface area, robust porosity, and structural stability, which makes them ideal for applications in storage devices. [17-18].

Conductivity and Structural Strength: When the temperature rises to $600\ °C$, the carbon atoms condense into stable & conjugated structures, which results in the formation of polycyclic aromatic hydrocarbons (PAHs) and aromatic rings similar to benzene. Also, when heated for a prolonged duration at higher temperatures of $700 - 900\ °C$, the disordered aromatic clusters combine and stack into graphitic domains [19].

The efficacies of the disordered graphitic clusters and aromatic nature are as follows:

> The delocalization of π-π electrons in the conjugated carbon network improves the electrical conductivity.

> The graphitic layers aromatic rings help in maintaining dimensional stability, even during thermal and electrochemical cycling.

Delocalized nature of the network prevents degradation of the electrode surface during repeated charge/discharge cycles.

Adsorption Capacity: The presence of polar functional groups, including hydroxyl (-OH), carboxyl (-COOH), and carbonyl ($>C=O$), markedly enriches the selective ion transfer behaviour, chemical reactivity and wettability of the material. The presence of polar functional groups, including hydroxyl (-OH), carboxyl (-COOH), and carbonyl ($>C=O$), markedly enriches the selective ion transfer behaviour, chemical reactivity and wettability of the material. These polar functional groups boost hydrophobicity, chemical reactivity and ion-exchange property. The pollutants in air or water are attached to these functional groups through hydrogen bonding or covalent bond formation [20].

 Depending on the pH, groups such as carboxyl and hydroxyl can ionize, providing active sites for cation or anion exchange. During the charge-discharge cycles, these oxygen-containing functional groups promote effective electrolyte ion adsorption and desorption by increasing the surface polarity, thereby providing active locations for ion interaction.

Applications for Electrochemical Energy Storage Devices

The superior structural and chemical properties, environmentally benign, renewable resources, and sustainable energy resource of biomass derived materials make it a productive asset for next generation energy systems. Through low-temperature and high-potential methods such as pyrolysis, chemical or physical activation rather than high temperature carbonization, highly ordered porous carbon is obtained. These carbon compounds have high electrical conductivity, thermal stability, and surface areas, making them ideal for application in many electrochemical energy storage devices. The organically occurring porous structure of biomass can be intensified to form micro-, meso-, and macropores, allowing for rapid ion transport and efficient electrolyte diffusion, both of which are critical for high energy and power density [21, 22].

Supercapacitors: Supercapacitors are energy storage devices that store energy through electrostatic charge accumulation by separating charges at the interface between an electrode and an electrolyte. Typically, supercapacitors are divided into two primary groups: i) Electrical Double-Layer Capacitors (EDLCs) and ii) Pseudocapacitors. Biomass-derived materials can be used as electrodes for supercapacitors owing to their porous nature, high thermal conductivity, and high carbon content. While reversible faradaic processes occur on the electrode surface in pseudocapacitors to store charges and energy, electrolyte ions are electrostatically deposited on the surfaces of the active electrode materials of EDLCs. In EDLC's type supercapacitors, typically the electrodes are carbonaceous materials due to their inherent properties of large surface areas, high conductivity, and superior physicochemical stability. In contrast, in pseudocapacitive type supercapacitors, metal oxides and polymeric composites are preferred as electrode materials to enhance the capacitance. In case of EDLC's the larger surface area is crucial in operation as the energy is stored by the accumulation of ions at the electrode-electrolyte interface. Also, the inherent porous structure allows easy diffusion of and rapid charge transport, thereby improving the power density of supercapacitors [23].

Furthermore, heteroatoms such as O, N, S, and P are often found in biomass materials, either naturally or as a result of synthesis by conversion techniques. During the reversible redox reactions, the heteroatoms are capable of forming surface functional groups such as -OH, -COOH, -NH$_2$, etc, which increase the total energy capacity, thereby contributing to pseudocapacitance [13]. To attain high energy density and power density collectively, a combination of hybrid capacitive performance of EDLC and pseudocapacitive behaviour. In addition to this, functionalization with other active groups enhances wettability property, which in turn enhances interaction with the electrolyte, resulting in better ion transport and diffusion.

The nature and origin have a considerable impact on the structural, chemical and electrochemical properties of biomass-derived materials significantly, which affects the performance of supercapacitors. The unique compositions of the various biomass resources, such as rice husk, coconut shell, sugarcane bagasse, wood waste, eggshell, and prawn shell, etc., lead to distinct properties in the carbon materials. In the case of rice husk, which is rich in silica and carbon, during activation process produces a highly ordered porous silicon carbon framework that acts as a medium for ion transport [2, 14]. The presence of silica helps maintain structural stability during the electrochemical process. [19].

Metal-ion batteries: Presently, energy storage systems rely significantly on metal-ion batteries, particularly those made of lithium, sodium, magnesium, and zinc. Their significance stems from the global drive toward high-performance, portable, and environmentally friendly energy solutions. The metal-ion batteries include Li-ion, Na-ion, Mg-ion, Zn-ion, etc., among which Li-ion battery rules the recent electronic market due to its high energy density and specific capacity [24]. The metal-ion batteries are composed of three main components such as electrodes (cathode & anode), electrolyte & separator. In case of Li-ion batteries, each component plays a crucial role in determining the energy density, capacity, stability, and recharge longevity. The commercially

Electrocatalysts and Advanced Materials for Sustainable Energy Storage   Materials Research Forum LLC
Materials Research Foundations 182 (2025) 160-173   https://doi.org/10.21741/9781644903797-11

available Li-ion batteries are constructed with $LiFePO_4$, $LiNiMnCoO_2$, $LiCoO_2$, etc. as cathode, graphite as anode, polyethylene or polypropylene as separator, $LiPF_6$ salt dissolved in propylene carbonate or ethylene carbonate or dimethyl carbonate as electrolyte. In the case of utilizing different materials, the energy density, capacity, and cycling stability of the batteries will differ according to the selection of the materials. For example, in case of $LiFePO_4$, the specific capacity is 160 mAh/g, and energy density is 90-140 Wh/kg, but for $LiNiMnCoO_2$ the specific capacity is 150-200 mAh/g, and energy density is 200-300 Wh/kg. Hence, selecting the suitable materials in each component plays a crucial role in determining the overall performance of the batteries. As best alternate and substitute to Li-ion batteries, Na-ion batteries emerge in line which offers better electrical conductivity and high energy density [25].

Both Li-ion and Na-ion batteries work in a similar principle and are composed of three main components: electrode, electrolyte, and separator. The electrolyte is usually Li-salt dissolved in an organic solvent, and polymeric materials, etc., the separator are ceramic coated, polymeric membranes, and inorganic solids, etc... In the case of electrode materials, the cathode is usually $LiCoO_2$, and the anode is graphite and hard carbon. The carbon atoms in the graphite are sp^2 hybridized and arranged in a layered hexagonal configuration. This unique arrangement and weak van der Waals forces between the layers make space for Li^+ ions to shuttle in and out without hinderance, aiding faster movement, resulting in long cycle life with better electrical conductivity. Hence, the Li^+ ions intercalate into the layers to form LiC_6 during charging. Conversely, during discharge the Li^+ ions will release from the in between layers, by this intercalation/de-intercalation takes place, making the battery performance with long cycle life. The graphite material is synthesized commercially through artificial synthesis methods such as high temperature thermal treatment (~2800–3000 °C). The treatment method consumes high energy, release CO_2, and relies heavily on non-renewable feedstocks such as petroleum based raw materials, and time-consuming processes [26].

To alternate and mimic the graphite structure, carbonized and partially graphitized carbon materials are derived from various plant- and animal-derived biomasses using different conversion techniques. Biomass-derived carbon materials exhibit properties similar to those of graphite materials, which are important in battery operation, including chemical and structural properties. The key chemical attribute of biomass-derived materials is the presence of heteroatoms which makes it best suitable for Li^+ ions adsorption and feasible electrolyte diffusion ensuing better conductivity and capacity. Considering structural properties, the high order porous structure with wide interlayer spacing accommodates more Li^+ ions. Hence, both the plant- and animal derived biomass materials provide the chemical and structural properties as discussed above.

Fuel Cells: A fuel cell is an energy conversion device that converts chemical energy to electrical energy through the continuous flow of fuel and oxidant, in which an oxygen reduction reaction occurs. Fuel cells consist of electrodes (cathode and anode), electrolyte, separators, catalysts, and bipolar plates. At the cathode, oxygen is reduced, and electrons flow to the other side through the external circuit. In contrast, protons (H^+) move through the electrolyte and combine with O_2 and electrons to form water; this is how the energy is converted. These components will differ in different types of fuels cells such as proton exchange membrane fuel cell, direct-methanol fuel cell, solid oxide fuel cell, alkaline fuel and microbial fuel cell [27] Electrolyte and separator are same in most fuel cells (i.e., proton exchange membranes), which makes it possible for electrons to transfer when the electrodes come into contact. The electrolytes in the main types of fuel cells include Nafion membranes, chitosan, and carbon quantum dots, wherein the electrodes include

Pt@C, Pt-Ru, and MnO_2, while the catalyst is a pervoskite material and separators such as graphene and graphene oxide. The primary rationale for selecting these materials is their structural, thermal, and chemical stability. In addition, when we look closely, we can find that the range of using other materials is limited by complex properties like structural integrity and inertness with reactants. When it is noted, carbon is used as such or as a support medium in all the components of the fuel cells. To achieve a better oxygen reduction reaction, the process should be selective for H^+ or OH^- ion transport, lower overpotential, and higher voltage. In the case of graphene oxide, the stacked interlayers of 0.6-1.2 nm allow only small ions, such as H^+ ions, and the presence of surface groups in GO when coordinated with water forms hydrated channels. These hydrated channels facilitate proton or hydroxide ion transport via the Grotthuss mechanism (proton hopping) or ion-solvent complex mechanism [28].

In addition, carbon derived from various biomass resources can be used as a supporting medium for noble metal nanoparticles (such as Pt or Pd) as electrodes/catalysts, improving dispersion and electron transfer. Owing to their renewable nature, high surface area, and tunable chemical composition, biomass-derived carbon materials doped with heteroatoms (such as N, S, P, and Fe) serve as non-precious metal catalysts (NPMCs) for the oxygen reduction reaction (ORR). Heteroatom-derived carbon from plant- and animal-based biomass acts as a metal-free ORR catalyst, especially in alkaline media. Furthermore, animal-derived biomass, chitosan, serves as a separator when processed into polymeric membranes, which also aids in selective ion transport. In addition, it is more flexible than commercially used graphite/graphene oxide, which is more rigid and requires additional chemical treatments. The Pt/C catalyst is very expensive, accounting for ~40–50% of the fuel cell cost, but the Fe–N–C catalyst from soybean husks is cheap as it is obtained from a natural source. In terms of durability, biomass-derived materials are stable in alkaline media for long-term operation, but commercial materials degrade with sintering and leaching. Comparably, the ORR activity is higher in alkaline media for biomass derived materials, conversely, high in acidic media for commercial materials. As a suitable carbon source derived from biomass (for example, Chitosan from prawn shells) as support media for catalysts, electrodes, and separators for fuel cell applications has attracted researchers because of its low cost, natural availability, and renewability. The conversion of available biomass into carbon for use in fuel cell applications as a support medium is a highly desirable value-added solution. Thus, biomass carbon represents a green and scalable solution for next-generation fuel cell technologies [29-30].

Comparative Analysis of Biomass derived material with commercial material

Biomass derived carbon materials and its derivatives are compared and its comparative analysis with the commercially available [1, 3, 8-9, 18, 24-25] are given in Table 4.

Table 4. Comparative analysis of various biomass derived material with the commercial materials used in various energy storage and conversion devices

Sl. No	Component	Biomass derived materials	Commercial materials	Comparative Analysis
1.	**Supercapacitors**			
i)	Electrode support	Activated carbon from wood, corn cob, buckwheat hull	Caron felt, graphene, carbon nanotubes	Biomass-derived carbon is cost-effective with high surface area
ii)	Binder	Lignin or starch-derived polymer	Polyvinylidene – difluoride (PVDF)	Green alternative to synthetic PVDF
2.	**Li-ion batteries**			
i)	Anode	Carbon from rice husk, banana peel, coconut shell	Graphite	Biomass has higher porosity & tunable structure; graphite has stable interlayers
ii)	Separator	Chitosan, Cellulose –	Celgard	Biomass derived are eco-friendly but lower mechanical stability; Celgard has higher mechanical stability
	Binder	Lignin, tanin based resins, chitosan blends	PVDF	Biomass derived are less toxic and renewable; commercial is more durable
3.	**Redox-flow batteries**			
i)	Cathode	Carbon porous electrode obtained from Buckwheat hull	Graphite felt	Biomass derived is cheaper and porous electrode with good redox kinetics
ii)	Separator	Biopolymers – Chitosan based	Nafion	Biomass derived are less toxic than synthetic; but lower conductivity than nafion
4.	**Fuel cells**			
i)	Catalyst	Biomass derived catalyst Fe-C-N from soyabean husk	Pt, IrO$_2$	Biomass catalysts are metal-free, durable in alkaline; Pt-based is expensive
ii)	Membrane	Chitosan based membrane derived from arthropod shells	Polymer membrane	Biodegradable

Challenges & Limitations

Although biomass-derived materials have numerous advantages, such as being naturally occurring, renewable, and ecologically benign, they also have several drawbacks about scalability, performance consistency, and material quality. A thorough understanding of the material synthesis methodology is necessary, even if difficulties are encountered during each conversion process, to overcome the limitations encountered [8]. Firstly, when taking into account of structural integrity, the biomass derived carbon obtained from different resources yields different properties. The growing environment and resource classification greatly influence the inherent properties, leading to differences in the porous nature, hierarchical structure, and other intricate properties [12]. This results in numerous problems, including scalability, inconsistency, and material quality. In addition, biomass-derived materials are inferior to commercial graphite and carbon-based materials in terms of conductivity. The as-derived materials require post-treatment to enhance conductivity. The presence of a highly ordered porous structure, which is important for redox reactions, is lacking. To enhance the ORR catalyst reaction, heteroatom doping is important, and metal nanoparticle inclusion is required. Therefore, the limitations and challenges of biomass-derived materials can be addressed by optimizing the material process and standardization, which is important for future industrial-scale applications.

Future Prospects

The urgent demand for sustainable, low-cost and environmentally friendly alternative materials to conventional materials make the future for the biomass derived material usage in energy storage and conversion devices. Along with the socio-environmental attributes, the biomass-derived materials are expected to make advancements in the areas of materials and surface designing and heteroatom doping, wherein the inclusion of N, S, & P will further enhance the electrochemical properties. Mainly, the materials suitable for all the electrochemical devices will be carbon, not only carbon also the composites and hybrid materials derived from carbon can be a suitable alternative to the conventional materials. In the line of composite and hybrid materials, to improve the structural, mechanical, thermal, chemical & catalytic properties, the biomass-derived carbon materials are combined with polymeric materials, 1D, 2D materials and metal oxides. Also, the future research should focus on other scalable green synthesis methods such as microwave-assisted, low temperature pyrolysis, template activation, and enzymatic activation, apart from the conventional pyrolysis, hydrothermal, and other methods, while considering the environmental impact.

The adaptability of biomass-derived bio-polymers such as chitosan, cellulose, and other synthetically produced bio-polymers from natural monomers such as polylactic acid, polybutyl succinate also open doors to the next generation of electrochemical devices. Their unique properties, including flexibility, and biocompatibility allow the biomass-derived bio-polymers to be utilised in wearable and flexible electronics. Besides, the nature and properties of the various biomass derived materials can be studied in detail theoretically with the help of the integration of AI & machine. These data driven technologies reduce experimental time, and energy thereby providing multiple conditions quickly. When it comes to a clean environment and circular economy, the utilization of biomass waste from agricultural and industrial sectors as energy resources not only supports sustainable development but also aid in providing carbon net zero. Anticipating forward, biomass-derived materials predicted to play an important role in emerging energy technologies such as solid-state systems, multi-valent batteries, CO_2 conversion, electrochemical water splitting, electrolyzers. Efforts must be put forward to enable commercialization over the biomass derived materials towards standardizing feedstocks, refining cost-performance feasibility, launching supportive policies, and life cycle assessment (LCA). Together, these progresses spot biomass-derived materials as cornerstones for the future of clean, sustainable and net-zero energy systems.

References

[1] Q. Hassan, P. Viktor, T. J. Al-Musawi, B. M. Ali, S. Algburi, H. M. Alzoubi, A. K. Al-Jiboory, A. Z. Sameen, H. M. Salman, M. Jaszczur, The renewable energy role in the global energy Transformations, Renew. Energy Focus. 46 (2024) 100545.. https://doi.org/10.1016/j.ref.2024.100545

[2] V. Solanki, S. Birman, Nature-Driven Renewable Energy Systems for Sustainable Development. In: P. Singh, P. Srivastava, A. Sorokin, (eds) Nature-Based Solutions in Achieving Sustainable Development Goals. Springer, Cham, 2024, pp. 131 - 166.. https://doi.org/10.1007/978-3-031-76128-7_5

[3] R. Jayabal, Towards a carbon-free society: Innovations in green energy for a sustainable future, 24 (2024) 103121.. https://doi.org/10.1016/j.rineng.2024.103121

[4] S. Thomas, M. Hosur, D. Pasquini, C. J. Chirayil (Eds), Handbook of Biomass, Springer Singapore, pp. XXVII, 1574.

[5] L. Zhang, Z. Liu, G. Cui, L. Chen, Biomass-derived materials for electrochemical energy storages, Prog. Polym. Sci. 43 (2015) 136 - 164.. https://doi.org/10.1016/j.progpolymsci.2014.09.003

[6] R. Millati, R. B. Cahyono, T. Ariyanto, I. N. Azzahrani, R. Utami Putri, M. J. Taherzadeh, Chapter 1 - Agricultural, Industrial, Municipal, and Forest Wastes: An Overview, M. J. Taherzadeh, K. Bolton, J. Wong, A. Pandey, (eds), Sustainable Resource Recovery and Zero Waste Approaches, Elsevier, 2019, pp. 1-22.. https://doi.org/10.1016/B978-0-444-64200-4.00001-3

[7] S. K. Tiwari, M. Bystrzejewski, A. D. Adhikari, A. Huczko, N. Wang, Methods for the conversion of biomass waste into value-added carbon nanomaterials: Recent progress and applications, Prog. Energy Combust. Sci. 92 (2022) 101023.. https://doi.org/10.1016/j.pecs.2022.101023

[8] A. Ephraim, P. Arlabosse, A. Nzihou, D. P. Minh, P. Sharrock, Biomass Categories. A. Nzihou, (eds) Handbook on Characterization of Biomass, Biowaste and Related By-products. Springer, Cham, 2020.. https://doi.org/10.1007/978-3-030-35020-8_1

[9] P. Adams, T. Bridgwater, A. L. Langton, A. Ross, I. Watson, Biomass conversion technologies. In: Greenhouse Gas Balances of Bioenergy Systems. Elsevier, Academic Press, 2018, pp. 107-139.. https://doi.org/10.1016/B978-0-08-101036-5.00008-2

[10] N. N. Nguyen, A. V. Nguyen, M. Konarova, Converting rice husk biomass into value-added materials for low-carbon economies: Current progress and prospect toward more sustainable

practices, J. Environ. Chem. Eng. 13 (2025) 115499..
https://doi.org/10.1016/j.jece.2025.115499

[11] D. Bisen, A. P. S. Chouhan a, M. Pant, S. Chakma, Advancement of thermochemical conversion and the potential of biomasses for production of clean energy: A review, Renew. Sustain. Energy Rev. 208 (2025), 115016.. https://doi.org/10.1016/j.rser.2024.115016

[12] S. K. Parakh, Z. Tian, J. Z. E. Wong, Y. W. Tong, From Microalgae to Bioenergy: Recent Advances in Biochemical Conversion Processes, Fermentation, 9 (2023) 529.. https://doi.org/10.3390/fermentation9060529

[13] B. Basak, R. Kumar, A.V. S. L. S. Bharadwaj, T. H. Kim, J. R. Kim, M. Jang, S. E. Oh, H-S. Roh, B-H. Jeon, Advances in physicochemical pretreatment strategies for lignocellulose biomass and their effectiveness in bioconversion for biofuel production, Bioresour. Technol. 369 (2023) 128413.. https://doi.org/10.1016/j.biortech.2022.128413

[14] X. Yang, Y. Zhang, P. Sun, C. Peng, A review on renewable energy: Conversion and utilization of biomass, Smart Molecules, 2 (2024) e20240019.. https://doi.org/10.1002/smo.20240019

[15] Y. Wang, M. Zhang, X. Shen, H. Wang, H. Wang, K. Xia, Z. Yin, Y. Zhang, Biomass-Derived Carbon Materials: Controllable Preparation and Versatile Applications, Small, 17 (2021) 2008079.. https://doi.org/10.1002/smll.202008079

[16] T. Khandaker, T. Islam, A. Nandi, Md. Al A. M. Anik, Md. S. Hossain, Md. K. Hasan, M. Hossain, Biomass-derived carbon materials for sustainable energy applications: A comprehensive review, Sustain. Energy Fuels, 9 (2025) 693-723.. https://doi.org/10.1039/D4SE01393J

[17] S. Ling, D. L. Kaplan, M. J. Buehler, Nanofibrils in nature and materials engineering, Nat. Rev. Mater. 3 (2018) 18016.. https://doi.org/10.1038/natrevmats.2018.16

[18] F. H. Isikgo, C. R. Becer, Lignocellulosic biomass: a sustainable platform for the production of bio-based chemicals .and polymers, 6 (2015) 4497-4559.. https://doi.org/10.1039/C5PY00263J

[19] M. Bartoli, M. Troiano, P. Giudicianni, D. Amato, M. Giorcelli, R. Solimene, A. Tagliaferro, Effect of heating rate and feedstock nature on electrical conductivity of biochar and biochar-based composites, Appl. Energy Combust. Sci. 12 (2022) 100089.. https://doi.org/10.1016/j.jaecs.2022.100089

[20] B. Li, D. Liu, D. Lin, X. Xie, S. Wang, H. Xu, J. Wang, Y. Huang, S. Zhang, X. Hu, Changes in Biochar Functional Groups and Its Reactivity after Volatile-Char Interactions during Biomass Pyrolysis, Energy & Fuels, 34 (2020) 14291-14299.. https://doi.org/10.1021/acs.energyfuels.0c03243

[21] Y-L. Bai, C-C. Zhang, F. Rong, Z-X. Guo, K-X. Wang, Biomass-Derived Carbon Materials for Electrochemical Energy Storage, Chem. Eur. J. 30 (2024) e202304157.. https://doi.org/10.1002/chem.202304157

[22] K. M. Darkwa, S. Akromah, R. M. Gupta, Chapter 9 - Advanced applications of biomass for energy storage, A. M. Mishra, C. M. Hussain, (eds), In Micro and Nano Technologies, Bio-

Based Nanomaterials, Elsevier, 2022, pp. 171-209,. https://doi.org/10.1016/B978-0-323-85148-0.00005-1

[23] W. Lu, Y. Si, C. Zhao, T. Chen, C. Li, C. Zhang, K. Wang, Biomass-derived carbon applications in the field of supercapacitors: Progress and prospects, Chem. Eng. J. 495 (2024) 153311.. https://doi.org/10.1016/j.cej.2024.153311

[24] C. Jin, J. Nai, O. Sheng, H. Yuan, W. Zhang, X. Tao, X. W. Lou, Biomass-based materials for green lithium secondary batteries, Energy Environ. Sci. 14 (2021) 1326-1379.. https://doi.org/10.1039/D0EE02848G

[25] A. Feng, X. Zhu, Y. Chen, P. Liu, F. Han, Y. Zu, X. Li, P. Bi, Functional Biomass-Derived Materials for the Development of Sustainable Batteries, ChemElectroChem, 13 (2024) e202400086.. https://doi.org/10.1002/celc.202400086

[26] A. Anwara, B. S. Mohammeda, M. A. Wahaba, M.S. Liew, Enhanced properties of cementitious composite tailored with graphene oxide nanomaterial - A review, Dev. Built Environ. 1 (2020) 100002.. https://doi.org/10.1016/j.dibe.2019.100002

[27] I. Sebbani, M. K. Ettouhami, M. Boulakhbar, Fuel cells: A technical, environmental, and economic outlook, Clean. Energy Syst. 10 (2025) 100168.. https://doi.org/10.1016/j.cles.2024.100168

[28] T. J. F. Day, U. W. Schmitt, G A. Voth, The Mechanism of Hydrated Proton Transport in Water, J. Am. Chem. Soc. 122 (2020) 12027-12028.. https://doi.org/10.1021/ja002506n

[29] J. Larmini, A. Dicks, Fuel Cell Systems Explained, Wiley, 2003, 2-s2.0-84949778632.X. https://doi.org/10.1002/9781118878330

[30] Ren, Y. Wang, A. Liu, Z. Zhang, Q. Lva, B. Liu, Current progress and performance improvement of Pt/C catalysts for fuel cells, J. Mater. Chem. A. 46 (2020) 24284-24306.. https://doi.org/10.1039/D0TA08312G

Keyword Index

About the Editors

Dr. Kalathiparambil Rajendra Pai Sunajadevi is a professor of chemistry at Christ University, Bangalore, India. She obtained her Master's degree in Chemistry from Mahatma Gandhi University, Kerala, and Ph.D. in Chemistry from the Cochin University of Science and Technology, Kochi, India. Her research focuses on diverse areas such as nanochemistry, organocatalysis, photocatalysis, materials science, metal organic frameworks, MXenes, energy materials, supercapacitors, and hydrogen evolution reactions through water splitting. Dr. Sunajadevi has made impactful contributions to the field of chemical sciences, with more than 125 publications in prestigious international journals, including those of the American Chemical Society, Royal Society of Chemistry, Elsevier, Springer, Wiley, and other reputed publishers. She has also authored several book chapters and holds patents for her innovative work in materials and catalysis. She is currently acting as reviewer for many of the journals and also there in the editorial team. She has guided several postgraduate and doctoral students throughout her academic career and actively participated in national and international conferences. Her research continues to support advancements in sustainable energy and green chemistry, fostering interdisciplinary collaboration and contributing to the global scientific community.

Dr. Dephan Pinheiro is currently working as a Professor at the Department of Chemistry, Christ University, (Deemed to be University), Bangalore, India. He completed his Master's in Chemistry from Bharathidasan University, Tiruchirappalli, in 1989, and his PhD from Christ University, (Deemed to be University), Bangalore. He has co-authored more than 70 research articles in high-impact journals and conference proceedings, besides contributing to several book chapters. He is passionate about using chemistry and materials research to address environmental challenges and is driven by a commitment to sustainability and the belief that science can create impactful solutions for a cleaner, greener future. His research interests include Nanochemistry, Heterogeneous Catalysis, Photocatalysis, Energy materials, and Materials Science.

Dr. Mothi Krishna Mohan serves as an Associate Professor in the Department of Sciences and Humanities, School of Engineering and Technology, Christ University, Bangalore. He completed his Ph.D. in Physical Chemistry from Cochin University of Science and Technology in 2014, supported by the prestigious UGC-BSR fellowship. He also holds a Master's degree in Inorganic Chemistry from Mahatma Gandhi University, Kottayam (2008). With over 12 years of teaching experience and 14 years of research expertise, Dr. Mohan's work concentrates on materials science. His research spans the development of heterogeneous catalysts for fine chemical synthesis, photocatalysts for environmental remediation, electrocatalysts for water splitting, and thin films for transparent conducting electrodes in device applications. He has contributed extensively to the field through peer-reviewed international publications, authored books, and chapters, and holds patents and copyrights. Dr. Mohan is well-versed in advanced chemical laboratory techniques and has managed both medium- and large-scale lab operations. In addition to his academic contributions, he serves as a reviewer for leading publishers such as Elsevier, Springer Nature, and Wiley. He is also an editor for *Scientific Reports*, a Q1 journal published by Springer Nature.

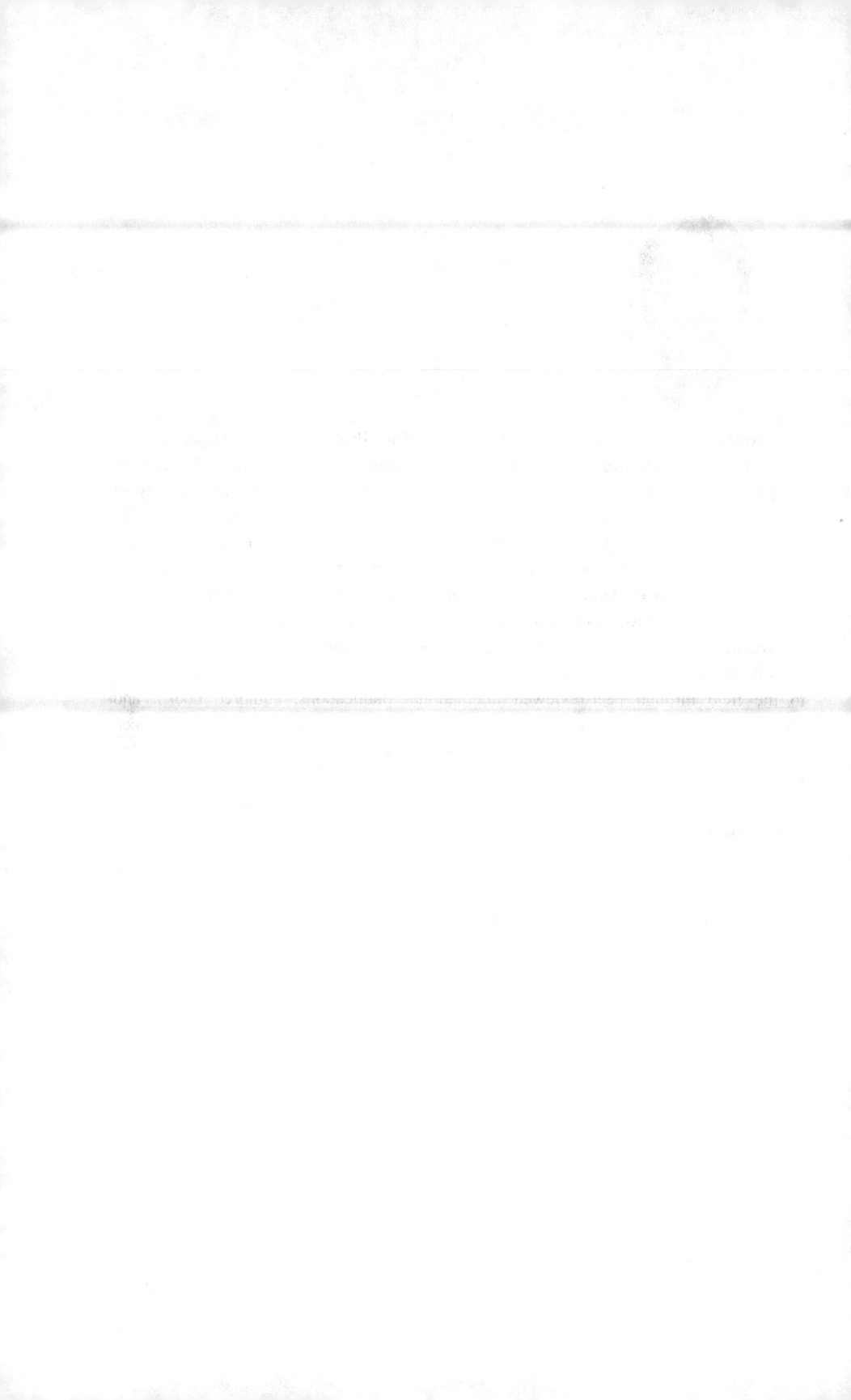

www.ingramcontent.com/pod-product-compliance
Lightning Source LLC
Chambersburg PA
CBHW071229210326
41597CB00016B/1998